T0302109

Lighting Upgrades

A Guide for Facility Managers

Second Edition

Lighting Upgrades

A Guide for Facility Managers

Second Edition

By Damon Wood, CLEP, LC

Routledge
Taylor & Francis Group

LONDON AND NEW YORK

Published 2020 by River Publishers

River Publishers
Alsbjergvej 10, 9260 Gistrup, Denmark
www.riverpublishers.com

Distributed exclusively by Routledge

4 Park Square, Milton Park, Abingdon, Oxon OX14 4RN
605 Third Avenue, New York, NY 10017, USA

Library of Congress Cataloging-in-Publication Data

Wood, Damon, 1956-
    Lighting upgrades : a guide for facility managers / Damon Wood.-
- 2nd ed.
        p. cm.
    Includes bibliographical references and index.
    ISBN 978-8-7702-2355-3 (print) -- ISBN 978-8-7702-2235-8 (electronic)
        1. Factories--Lighting. 2. Factories--Energy conservation. 3. Elec-
tric lighting. 1. Title.

    TK4399.F2W69 2004
    658.2'4--dc22

                                                            2003064285

Lighting upgrades : a guide for facility managers, second edition/Damon Wood.
First published by Fairmont Press in 2004.

Routledge is an imprint of the Taylor & Francis Group, an informa business

0-88173-329-6 (The Fairmont Press, Inc.)
978-8-7702-2355-3 (print)
978-8-7702-2235-8 (online)
978-1-0031-6926-0 (ebook master)

iv

Dedication

*To my wife, Debbie, and my kids, Sara and Kevin,
who generously allowed me to invest hundreds of hours
of "free time" in writing this book.*

Contents

Foreword

Today's facilities manager must be a "jack of all trades," informed on every issue that impacts the workplace and its occupants, and able to handle each day's crises on the spot! Certainly, those in this role know that these are unrealistic goals at best—and impossible, at worst.

This book, *Lighting Upgrades: A Guide For Facility Managers*, by Damon Wood is an excellent "crash course" on all aspects of lighting in the workplace. Professionals with more than a passing knowledge of lighting must have *Lighting Upgrades* on their reference bookshelf. It is an extremely well-organized, illustrated and comprehensive guide.

For those managing a monastery whose occupants insist upon the continued use of candles as their sole light source, I suggest you read no further. For the rest of us, however, I believe that *Lighting Upgrades* is mandatory reading. That includes *anyone* associated with the interior or exterior of buildings, such as facilities managers, architects, building owners, contractors and interior designers. It is imperative that all of us adequately understand the principles of lighting and light's impact on conservation, productivity, and safety! *Lighting Upgrades* addresses all of those issues in lay terms and in a logical fashion.

I shall say no more in order for readers to discover themselves what I have learned. That is, how little most of us have understood about this subject until reading *Lighting Upgrades: A Guide for Facility Managers*. Thank you, Damon Wood!

Kreon L. Cyros, Director
Office of Facilities Management Systems
Massachusetts Institute of Technology

Introduction

WELCOME

As new lighting technologies are introduced at increasing rates, building owners now have unprecedented opportunities for reducing operating costs while enhancing aesthetics, safety, and productivity. But as the number of lighting upgrade options expands, facility managers struggle to stay current with the latest innovations. Besieged by vendors with conflicting performance claims, facility managers are urgently seeking unbiased application guidance. To take full advantage of these opportunities, current lighting needs must be carefully assessed, and lighting product performance must be objectively evaluated. *Lighting Upgrades: A Guide For Facility Managers* is designed to equip facility managers with the knowledge and resources to reap the bountiful rewards of energy-efficient lighting. This book is structured to provide readers with rapid access to the most useful information:

LIGHTING FUNDAMENTALS

The first two chapters present the fundamentals of lighting quality and efficiency that apply to all lighting upgrade applications. The emphasis on lighting quality and efficiency is established in these chapters and is evident throughout the book.

LIGHTING SYSTEMS AND UPGRADES

Chapters 3-10 address the most common varieties of commercial lighting equipment and their upgrade options. The *odd* chapters in this series (Chapters 3, 5, 7, 9) focus on the baseline technologies used in fluorescent, compact-size, high-intensity discharge and emergency lighting systems. The *even* chapters (Chapters 4, 6, 8, 10) provide specific guidance for applying new lighting upgrade technologies in these systems.

CONTROLS

Chapters 11 and 12 help readers gain control over their lighting systems by identifying cost-effective applications of automatic switching and dimming systems. The use of lighting controls can help minimize energy waste while offering users greater flexibility in meeting their lighting needs.

APPLICATIONS

Chapters 13-17 take the concepts and guidelines presented in earlier chapters and apply them to specific lighting applications such as office, retail, industrial, and outdoor lighting. The practical guidelines offered in these chapters address the application-specific lighting upgrade objectives and technologies to consider.

GETTING THE JOB DONE

Chapters 18-20 guide the reader through the steps of planning, implementing and maintaining lighting system upgrades. These steps include lighting surveys, analysis, trial installations, financing, project management, lighting system maintenance, and waste disposal.

REFERENCE INFORMATION

The final sections of the book provide valuable reference information, including trade and professional associations, federal government resources, lighting-related periodicals, a bibliography, and a glossary of terms.

THE BENEFITS

After reading this book, there is no doubt that you will begin to see your lighting systems in a whole new "light." You will no longer view them as a drain on corporate resources; they will represent a potential profit center that can turn a modest investment into improved cash flow and enhanced worker productivity.

Chapter 1

The Elements of Lighting Quality

Maintaining high standards for lighting quality is one of the highest priorities in facilities management. High quality lighting can enhance safety, productivity and aesthetics. Lighting upgrades that improve lighting quality will deliver the right amount of light, maintain uniform illumination, control glare and improve color appearances. This chapter provides guidance in selecting technologies that improve lighting quality.

OVERVIEW OF LIGHTING QUALITY MEASURES

In a typical business, people costs (salaries and benefits) can be more than 100 times greater than the costs of operating the lighting system! Although it is difficult to quantify, there is a direct relationship between the quality of the lighting system and worker productivity. Therefore, any increase in productivity that is caused by an improved lighting system will have a profound effect on profits. Conversely, a lighting modification that reduces productivity—even by as little as one percent—could destroy the project's profitability.

To specify the components of a high quality lighting upgrade, many of the qualitative attributes of the lighting system must be defined. This section provides a brief overview of the most common measures of lighting quality. Following this overview, more detailed discussions of these measures are provided.

Illumination Levels

Modifying lighting systems to provide the proper *quantity* of light can contribute to improved lighting quality. When deciding on light levels, "more" does not necessarily mean "better." All lighting upgrades

1

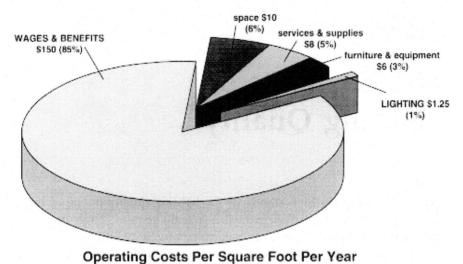

Operating Costs Per Square Foot Per Year
(Typical Office Space)

Figure 1-1. Because "people costs" are so much greater than lighting costs, lighting upgrade decisions should be driven by lighting quality considerations and their potential impact on productivity. Courtesy: EPA Green Lights.

should be designed to deliver the light levels that are appropriate for the specific visual task(s) being performed in the space. Existing light levels may need to be increased or decreased to meet the revised target light level—and improve lighting quality. Illumination levels are measured in *footcandles* (fc).

Uniformity of Illumination
        Modifications to light fixtures—more appropriately termed within the industry as "luminaires"—can result in reductions in the uniformity of light levels across the workplane. The workplane is an "imaginary" plane resting at the level where the primary visual tasks are located. The workplane (and respective task) can be horizontal—such as writing at a desk, or vertical—such as reading the spines on books on library shelves. Distraction, visual fatigue and unacceptable levels of illumination can result from reduced uniformity.

Visual Comfort
        Productivity can suffer if glare from the luminaires causes reductions in contrast on work surfaces or computer screens. By specifying

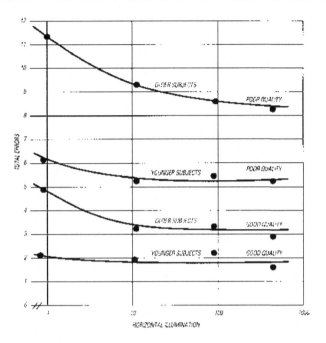

Figure 1-2. The results of a test conducted by the Illuminating Engineering Research Institute (IERI) showed that improving lighting quality (color rendering and visual comfort) had a greater impact on reducing errors than simply increasing average light levels (10-100 fc). The test's visual task was proofreading mimeographed documents. *Courtesy: National Lighting Bureau.*

upgrade components that provide improved glare shielding, workers can experience gains in productivity and reductions in fatigue and eye strain. Also, workstations can be reoriented to minimize direct glare from windows and reflected glare from luminaires. In office environments, visual comfort and performance can be further improved by changing the distribution of the lighting so that it brightens walls and ceilings, which reduces the range of surface brightness levels within the user's field of view.

Color Rendering

The ability to see true colors is another product of good lighting quality. Light sources vary in their ability to accurately reflect the true colors of people and objects. Choosing sources with high color rendering values will make colors appear brighter and more natural, improving the appearance of the space. Occupants may perceive that the space is more brightly lighted when the color rendering is improved.

Color Temperature

A lamp's color temperature indicates the "warmth" or "coolness" of its light. Contributing to the ambiance in an illuminated space, color temperature is an architectural choice. It is usually considered desirable to maintain a uniform color temperature throughout a given space, although for certain purposes warm and cool lamps can work together well.

METHODS FOR IMPROVING LIGHTING QUALITY

There are a number of methods available to facility managers and practitioners of lighting services to improve lighting quality in existing spaces, or ensure that the desired high levels of lighting quality are achieved in new spaces. In this section, we will review specification of light levels, ensuring uniform distribution of light, controlling glare, improving visual comfort and improving color.

Improving Quality With Corrected Light Levels

The first question to ask when considering any lighting system modification is, "How much light is needed for workers to perform the visual tasks in this space?" Unfortunately, too many lighting upgrade decisions are made without addressing this critical question. By assuming that the goal is to "maintain" the current light level, specifiers may fail to correct underlighted or overlighted conditions, thereby perpetuating a compromised level of productivity. In some cases, the visual tasks may have changed since the lighting system was first designed. It may make sense to *reduce* light levels to a new recommended level, thereby saving additional energy dollars while promoting productivity and user acceptance.

Selecting Target Illumination Levels

The Illuminating Engineering Society of North America (IESNA) has simplified the procedure for determining the appropriate average light level (illuminance) for a particular space. Table 1-1 lists seven generic categories of visual tasks in order of increasing difficulty. Each category is assigned a letter designation (A through G) and a corresponding footcandle value. In addition to these generic descriptions, detailed listings of Illuminance Categories for more than 600 specific visual tasks are provided in the current edition of the IESNA *Lighting Handbook* and related IESNA publications.

Table 1-1. Illuminance recommendations.
Source: Illuminating Engineering Society of North America.

Illuminance Category	Recommended Illuminance
A - Public spaces	3 fc
B - Simple orientation for short visits	5 fc
C - Working spaces where simple visual tasks are performed	10 fc
Performing visual tasks of:	
D - High contrast and large size	30 fc
E - High contrast and small size, or low contrast and large size	50 fc
F - Low contrast and small size	100 fc
G - Near threshold visibility	300-1000 fc

Lighting Upgrades for Modifying Light Levels
Once the target footcandle levels are selected, the appropriate lighting upgrades are specified based on:

- Luminaire efficiency and light distribution.
- Lamp lumen output.
- The effects of light losses from lamp lumen depreciation and dirt accumulation (see Chapter 20).
- Room size and shape.
- Availability of natural light (daylight).

An entire chapter of this book has been devoted to the subject of modifying light levels based on revised light level requirements. Refer to

Chapter 13 for guidance in selecting technologies for modifying light levels.

Improving Quality with Uniform Illumination
When a lighting system is specified during the building design phase, designers space the overhead luminaires to provide the right distribution of light from the luminaires so as to gain uniform illumination across the workplane. This is accomplished by keeping the horizontal spacing between luminaires within the maximum distance recommended by the manufacturer.

Spacing Criterion
Published in the luminaire's photometric report is a value known as the *spacing criterion*—a number generally between 0.5 and 2.0. To calculate the maximum recommended horizontal spacing between luminaires (to ensure uniformity), multiply the spacing criterion by the distance between the luminaire and the work surface. Note that if a luminaire has a non-symmetrical light distribution pattern, it will have two different spacing criteria—one for spacing luminaires in a direction perpendicular to the lamps and one for spacing the luminaires in a direction parallel to the lamps.

Potential Effects of Lighting Retrofits on Uniformity
Modifications to existing luminaires can cause reductions in lighting uniformity. The problems that non-uniform illuminance causes include:

•   Inadequate light levels in some areas.

•   Visual discomfort or distraction when an occupant's visual tasks are spread across overlighted and underlighted areas.

•   Low-quality appearance resulting from shadows on walls and patches of light on floors.

For example, retrofit low-glare louvers (designed to replace clear lenses on fluorescent luminaires) will reduce uniformity by shielding light at higher angles, directing more of the light straight down. As a result, the light level between luminaires may be reduce to unacceptably low levels.

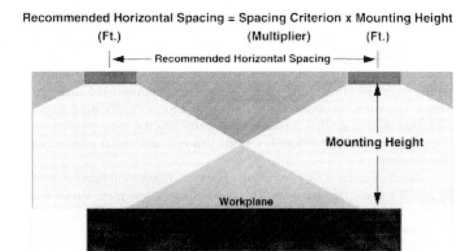

Figure 1-3. The spacing criterion from the luminaire's photometric report is used to determine the recommended maximum horizontal spacing between luminaires to ensure uniform illumination on the work surface. *Courtesy: EPA Green Lights.*

Specular reflectors are another common lighting retrofit that can reduce uniformity. Although advanced designs can maintain the existing luminaire's spacing criterion, the specular (mirrorlike) nature of these reflectors tends to direct more of the light in a downward direction, reducing lighting uniformity.

In most cases, measures for reducing glare (louvers and reflectors) will involve some sacrifice in uniformity, unless the luminaires are moved closer together. Ask suppliers of louvers or reflectors to install trial installations for evaluating the effects on lighting uniformity.

Improving Quality by Controlling Glare

Glare from improperly shielded light sources can have detrimental effects on worker productivity. In general, luminaire light output at high angles (closer to horizontal) can cause the greatest visual discomfort. The primary problem with high-angle brightness is the distracting (and sometimes disabling) reflected glare that obscures the images on computer screens.

Methods for Controlling Glare

Glare can be controlled by using improved lighting equipment that is specifically designed to reduce glare. A louver is commonly used to block occupants' direct view of a light source. By specifying a louver with a high shielding angle, visual comfort can be improved. Refer to Figure 1-4 which illustrates the definition of a louver's shielding angle. Note that when a shallow louver is used (assuming the same "cell" size), the shielding angle is reduced, resulting in reduced visual comfort or glare control.

Another method for controlling glare from fluorescent luminaires is to install a low-glare flat lens. The most efficient low-glare lenses are clear, utilizing an optical design to direct light in a more downward direction. Tinted lenses may improve visual comfort, but they can trap most of the light within the luminaire.

Indirect lighting—or uplighting—can create a low-glare environment by uniformly lighting the ceiling. However, if the uplighting system concentrates light on the ceiling directly above the luminaire, the patterns of ceiling brightness can distract workers. By selecting indirect luminaires with the proper wide-angle light distribution, and by suspending the luminaires at the proper distance, the indirect lighting system can provide a high quality, productive working environment.

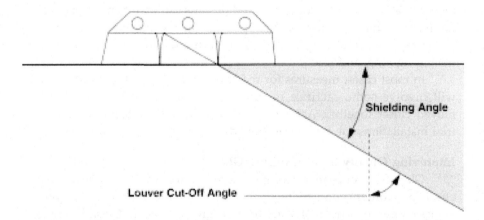

Figure 1-4. Light emitted from luminaires at higher angles is more likely to cause glare. Louvers with greater shielding angles provide improved visual comfort. *Courtesy: EPA Green Lights.*

Reflected glare on horizontal surfaces is another problem faced by workers. When publications are printed on glossy materials, this problem becomes particularly annoying. The most direct method for dealing with this form of reflected glare is to either reposition the visual task or reposition the offending luminaires. In some cases, a polarizing lens can reduce reflected glare and improve contrast on horizontally oriented visual tasks.

Improving Quality with High Color Rendering Lamps

Lamps with improved color rendering performance can improve aesthetics and boost worker productivity where color-based tasks are performed. In addition, improved color rendering can increase occupant perceptions of room light levels.

The Color Rendering Index (CRI)

The color rendering index (CRI) scale is used to measure how "accurately" object colors will appear under the light source. The CRI is defined on a scale between 0 and 100, with a higher CRI indicating better color rendering, or less color shift from "true." CRIs in the range of 75-100 are considered "excellent," 65-75 are "good," 55-65 are "fair," and 0-55 are "poor."

Typical CRI Values for Common Light Sources

The *typical* CRI values for selected light sources are listed in Table 1-3. Some manufacturers offer special lamps with extraordinarily high color rendering values.

Improving Quality with Appropriate Color Temperature Lamps

The color temperature of a lamp measures the "warmth" or "coolness" of its light, which can have a significant effect on the "ambiance" of the illuminated space. Color temperature is quantified as the absolute temperature of a blackbody radiator, expressed in degrees Kelvin (K). A blackbody radiator—a laboratory device made of a heat-conducting material such as iron—changes color as it is heated to extremely high temperatures, first to red, then to orange, yellow and finally bluish-white at the highest temperature. *Note that a "warm" color light source actually has a lower color temperature.* For example, a cool-white fluorescent lamp appears bluish in color with a color temperature of around 4100K. Table 1-4 indicates the general categories of color temperature.

Table 1-3. Typical CRI values for common lamp types.

Source	Source Lamp Type	Typical CRI Values
INCANDESCENT/HALOGEN	Incandescent/Halogen	100
FLUORESCENT	Cool White T12	62
	Warm White T12	53
	Triphosphor T12	73-90+
	T8	75-90+
	T10	82-90+
	Compact Fluorescent	82-86
MERCURY VAPOR	Clear Lamps	15
	Coated Lamps	50
METAL HALIDE	Clear Lamps	65
	Coated Lamps	70
	Ceramic Arc Tube	83+
HIGH-PRESSURE SODIUM	Standard	22
	Deluxe	60-65
	White HPS	70-85
LOW-PRESSURE SODIUM	Low-pressure Sodium	0

Table 1-4: General categories of color temperature.

General Category	Color Temperature Range
"Warm" (yellow-white)	Less than 3500K
"Neutral" (white)	3500-4000K
"Cool" (blue-white)	More than 4000K

Typical Color Temperature Values for Common Light Sources

The diagram in Figure 1-6 illustrates the typical color temperature of a variety of light sources. Note that with the use of various phosphor coatings, fluorescent lamps can produce a wide range of color temperatures—from less than 3000K to over 5000K. In addition, new metal halide lamp designs can now produce color temperatures as low as 3000K, approaching the look of incandescent halogen lighting.

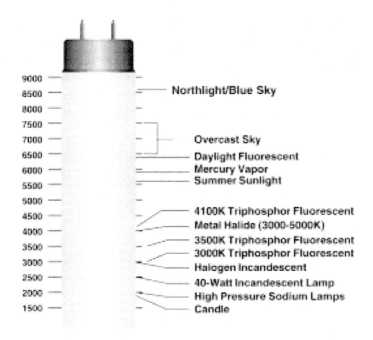

Figure 1-5. **Typical color temperature values for common light sources.**
Courtesy: EPA Green Lights.

Finding the Appropriate Color Temperature

The selection of color temperature is an architectural consideration and should therefore be based on "taste" or "preference" rather than on a strict metric. As a general rule, designers will specify warmer color temperatures in spaces with lower illumination levels, such as in intimate dining areas. Conversely, high-illumination spaces such as drug stores are normally lighted with cooler color temperature sources. In other words, environments with increased activity will normally feature increased illuminance and higher color temperatures. Where neutral-white lighting is desired, color temperatures of 3500-4000K should be specified.

Because changes in color temperature are more noticeable to occupants than changes in color rendering, lighting upgrades should limit changes in color temperature to less than 600K. In addition, if multiple lamp types are used in the same space (such as compact fluorescent task lighting and linear fluorescent ambient lighting), all light sources should have the same color temperature.

For practical purposes, there is no correlation between color temperature and color rendering. For example, a light source with a low color temperature ("warm" source) is not necessarily a "high quality" source. Although both incandescent (CRI = 100) and low-pressure sodium lamps (CRI = 0) are considered warm sources, they are opposites in terms of color rendering performance. *However, to objectively compare color rendering performance, evaluate lamps with the same color temperature.*

Chapter 2

Maximizing Lighting Efficiency

When selecting energy-efficient lighting upgrade equipment, the goal is to determine the correct mix of technologies that will deliver the desired quantity and quality of light using the least amount of energy. Efficient lighting upgrades should address the light sources (lamps and ballasts), the optical components (the light fixture, or luminaire) and the automatic control of lighting operation.

LIGHTING EFFICIENCY METRICS

Before we can explore the various strategies for reducing energy use, we must first have a clear understanding of how lighting efficiency is measured. We will begin by defining the units of measurement used as the building blocks for quantifying and assessing lighting efficiency.

Power Input: Watts (W) and Kilowatts (kW)

The watt is a unit for measuring electrical power. It defines the rate of energy consumption by an electrical device. The power input to a luminaire is usually measured in watts (W). However, the total power requirements for a series of luminaires—a lighting system—is usually defined in terms of kilowatts (kW). (One kilowatt equals 1,000 watts; refer to the glossary in Appendix II for complete definitions of these and other electrical terms.) Most commercial and industrial electricity rates include a charge for electricity demand—measured in kW—which can represent a significant portion of the electricity bill.

In lighting circuits, power input is related to voltage and amps by the formula:

$$\text{Watts} = \text{Volts} \times \text{Amps} \times \text{Power Factor}$$

Wattages for lighting products are published in the manufacturer's catalog, which are typically based on tests performed in controlled laboratory conditions as specified by the American National Standards Institute (ANSI). However, specific application conditions may cause some variation in actual wattage. If the most realistic data is required, input wattage can be measured in the field using a digital watt-meter that integrates measurements of volts, amps and power factor.

Energy Consumption: Kilowatt-hours (kWh)
The energy consumed by an electrical device is the wattage of the device multiplied by its hours of use. For example, a kilowatt-hour (kWh) of electricity consumption results when a 1000W (1 kW) device is operated for one hour, or a 500W device operates for two hours, and so on. It is the product of input wattage (in kW) and operating time (in hours per year), expressed:

Energy Consumed	=	Power	×	Time
Energy Consumed	=	Input Wattage	×	Operating Hours
kWh	=	kW	×	h

A kWh figure can be determined for any given period of time, although the most meaningful calculations base the operating hours on annual operating hours so that the resulting kWh figure is the amount of energy consumed in one year of operation.

It is important to carefully estimate the annual operating hours of lighting systems to determine annual energy consumption and resultant costs. Exit signs are easy, as they operate 24 hours per day, 365 days per year, which is 8,760 hours. To develop more reliable estimates of operating hours for other lighting systems, consider using the simple time-logging devices that attach magnetically onto the inside of luminaires; when the lights are turned on, the logging device's photo sensor activates a clock that records the hours of actual lighting operation.

Therefore, to fully evaluate options for reducing energy consumption (and costs), consider technologies that reduce *hours of operation* as well as those that reduce *input watts*.

Lamp Light Output: Lumens (lm)
The total light output of a lamp is measured in lumens. The lumen is a measure of light flow or luminous flux. Note that as lamps age, their

light output decreases (i.e., lamp lumen depreciation occurs). Most lamp ratings are based on "initial" lumens—the lumen output measured after the first 100 hours of lamp operation. Lamp catalogs also list the "design" lumens—the lamp's lumen output at 40 percent of its rated life—which indicate the lamp's *average* lumen output over its life.

Luminaire Light Output: Lumens (lm)

The total light output from a luminaire is also measured in lumens. However, because the luminaire usually absorbs a portion of the light produced by the lamps, the luminaire's output is less than the total lumen output from the lamp(s) inside the luminaire (see luminaire efficiency below). Note that as the luminaire becomes dirty or its optical surfaces deteriorate, the luminaire's lumen output will decrease.

Workplane Illumination: Footcandles (fc)

One of the most important units of lighting measurement is the footcandle, which measures the light intensity on a plane at a specific location. This light intensity is referred to as *illuminance* or light level. Illuminance is measured in footcandles, which is an expression of lumens per square foot on the workplane. The *footcandle* is the English unit of illuminance measurement; *lux* is the metric unit for illuminance, measured in lumens per square meter. To convert footcandles to lux, multiply footcandles by 10.76. Illuminance can be measured using a light meter located on the work surface where tasks are performed.

Note, however, that we cannot actually see a footcandle; what we see is the effect of footcandles reflecting off of surfaces. This effect is commonly described as "brightness"—more technically referred to as "luminance"—and is measured in *footlamberts* (English units) or *candelas per square meter* (metric units).

Putting It All Together: Expressing Lighting System Efficiency

The process of converting electrical power to workplane illumination is diagrammed in Figure 2-2. Using the units of measurement described above, the efficiency of various aspects of the lighting system's performance can be evaluated. The following measures of lighting system efficiency can be applied to the various steps in this process.

Source Efficacy (lm/W or LPW)

Source efficacy refers to how efficiently a light source (or lamp-and-

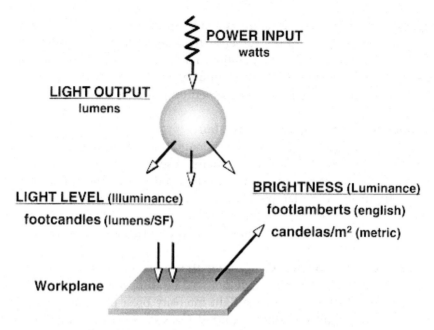

Figure 2-1. The units of measurement typically utilized in lighting efficiency evaluations include watts, lumens and footcandles. *Courtesy: EPA Green Lights.*

ballast system) converts power input (watts) to light output (lumens), and is therefore measured in lumens per watt (lm/W or LPW). The term *source efficacy* is the "miles-per-gallon" metric for evaluating light source fuel efficiency. Clearly, the light sources that produce high values of efficacy should be considered for lighting upgrades. However, the other measures listed below should also be considered.

Luminaire Efficiency
[Luminaire Lumens ÷ Bare Lamp Lumens, (%)]
    Luminaire efficiency is defined as the percentage of light produced within a luminaire that actually leaves the luminaire, rather than being absorbed inside.
    For example, if two luminaires use the same number and types of lamps and ballasts, more light will be emitted from the luminaire with the higher efficiency. One goal of efficient illumination is to select luminaire components (lenses, louvers, reflectors) that improve luminaire

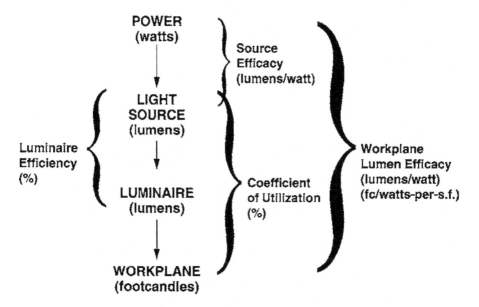

Figure 2-2. There are several measures that describe the efficiency of one or more steps in the process of converting incoming electricity (wattage) into footcandles on the workplane. *Courtesy: EPA Green Lights.*

efficiency by absorbing a lower percentage of the produced light. Note, however, that it may be necessary to sacrifice maximum luminaire efficiency to achieve desired improvements in visual comfort.

Coefficient of Utilization
[Workplane Lumens ÷ Bare Lamp Lumens, (%)]

Although the values for source efficacy and luminaire efficiency are useful for comparing the performance of similar technologies, neither of these two measures make reference to the system's ability to deliver light to work surfaces. The coefficient of utilization (CU) takes into account the luminaire efficiency, light distribution pattern, room geometry, luminaire mounting height, task height and room surface reflectances. A lighting specifier uses the CU to predict the footcandles that will be achieved by a specific lighting system mounted in a defined room.

The CU value is found in a table published in the luminaire's photometric report. In addition, most lighting software programs include databases of photometric information for specific luminaires. The appro-

LUMINAIRE

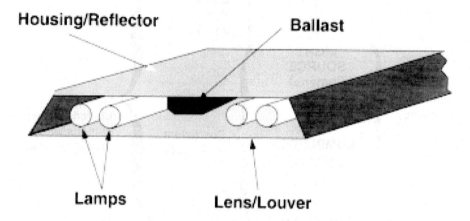

Figure 2-3. A typical fluorescent light fixture and its components. Note that the technical term for a light fixture is "luminaire." *Courtesy: EPA Green Lights.*

priate CU value is found in the table based on the calculated "room cavity ratio" and the room surface reflectances. The room geometry, luminaire height and task height are represented by the room cavity ratio, and the room reflectances are selected for the ceiling, walls and floor. Refer to Chapter 18 for information on how to use the CU values for predicting illumination performance of lighting upgrades.

Workplane Lumen Efficacy
[Workplane Lumens ÷ Input Power, (lm/W)]

Similar to light source efficacy, workplane lumen efficacy is expressed in lumens per watt. However, this measure indicates the quantity of lumens falling on the workplane for each watt of power applied. To calculate the workplane lumen efficacy, divide the average illuminance (fc) in the room by the lighting system's power density (W/sq.ft.). When dividing footcandles by watts-per-square-foot, these units convert to lumens per watt (because a footcandle is a lumen per square foot). This measure may be most meaningful when comparing different types of proposed lighting systems. It measures the overall objective of energy-efficient lighting design—to deliver the target footcandle level using the least watts.

LIGHT SOURCE EFFICACY
(maintained)

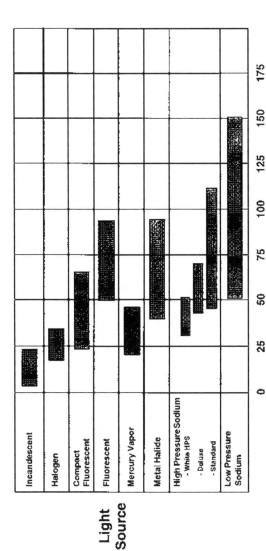

NOTE: Source efficacy values include ballast losses and lamp lumen depreciation at 40% lamp life.

Figure 2-4. Within each light source family, there are wide variations in how efficiently light sources convert electricity into light. Typically, the highest efficacy performance is achieved in higher-lumen sources and/or when electronic ballasts are used. See Chapters 3, 5 and 7 for more detailed information about fluorescent, compact fluorescent and high-intensity discharge lighting. *Courtesy: EPA Green Lights.*

Unit Power Density (Watts Per Square Foot)

Calculating the unit power density provides an indication of lighting system efficiency, when compared to other spaces with similar room geometry and light level requirements. For example, high-quality office lighting can now be delivered for less than one watt per square foot.

Energy Utilization Index (kWh Per Sq.Ft.)

This measure indicates the intensity of energy usage per square foot of illuminated floorspace. Unlike the unit power density measurement, this index includes the *dimension of time*. Therefore, spaces with 24-hour-per-day operation may yield a higher energy utilization index for lighting. Conversely, the use of automatic controls to turn off lights may yield a lower energy utilization index.

MAXIMIZING ENERGY EFFICIENCY

Although there are numerous issues to consider with any lighting upgrade project, there are three principles that should be applied to *all* lighting upgrade projects. To maximize energy savings, follow these important guidelines:

1. *Determine the light level that is needed for occupants to perform their specific visual tasks.* When investigating lighting upgrades, it is easy to ignore this important principle. The temptation is to simply assume that the existing light levels are adequate. In many cases, however, the visual tasks in the space may have changed since the lighting system was originally designed, and the light levels should not remain the same. Reducing light levels in overlighted spaces can be a quite cost-effective way of saving energy while maintaining or improving lighting quality. Refer to Chapter 13 for a complete discussion of upgrade technologies used for modifying light levels.

2. *Efficiently produce the light and deliver the target illumination to the visual task.* In other words, select the combination of energy-efficient lamps, ballasts, lenses/louvers and other components that will deliver the right amount (and quality) of light using the least watts.

Table 2-1. Applications of lighting efficiency measures.

Measure	Application
Source Efficacy	To select lamp and ballast types.
Luminaire Efficiency	To select lenses, louvers, reflectors, new luminaires.
Coefficient of Utilization	To specify new luminaires; to calculate footcandles.
Workplane Lumen Efficacy	To compare overall performance between dissimilar lighting systems.
Unit Power Density	To screen facilities for lighting efficiency.
Energy Utilization Index	To screen facilities for lighting efficiency and controls.

Again, do not "jump the gun" and look for least-wattage technologies that will simply maintain the existing light level if, in fact, the light level requirements differ from what is currently provided by the lighting systems.

3. *Automatically control the lighting operation.* Minimizing wattage is only part of the goal in achieving energy savings. Eliminating unnecessary lighting operation through occupancy sensors, scheduling systems and daylight controls can also achieve significant energy savings.

To maximize energy savings (and profit), *each* lighting upgrade project should address *all* of these critical principles.

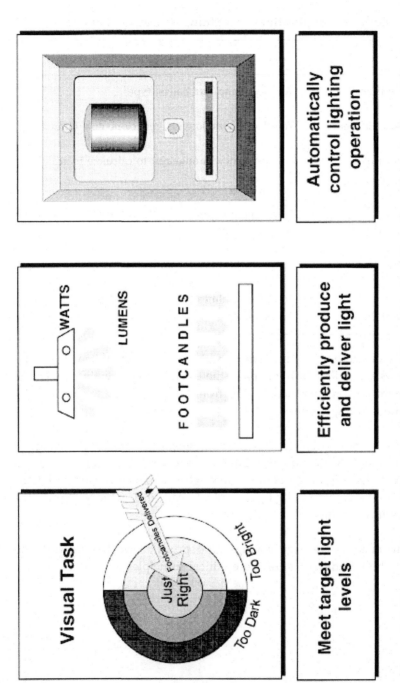

Figure 2-5. To maximize energy savings in *any* lighting upgrade, follow these three fundamental principles. *Courtesy: EPA Green Lights.*

Chapter 3

Full-size
Fluorescent Equipment

F luorescent lighting is the most commonly used commercial light
source in North America, illuminating over 70 percent of commercial building space in the United States alone. Because most new
fluorescent technologies are simply variations of existing products, it is
important to understand the components of existing fluorescent lighting
systems. Here, we will focus on "full-size" fluorescent lamps, ballasts
and luminaires.

FULL-SIZE FLUORESCENT LAMPS

Full-size fluorescent lamps are produced in both linear and U-
shaped configurations, with diameters ranging from 5/8 inches to over
2 inches. Because fluorescent lamps are linear light sources (as opposed
to concentrated "point" light sources), it is difficult to direct fluorescent
light in a well-controlled beam. The diffuse quality of fluorescent light
generally requires that fluorescent systems be used in applications with
mounting heights of less than 30 ft. These applications include general
ambient lighting, wall washing and sign lighting.

Types of Full-size Fluorescent Lamps

Full-size fluorescent lamps are categorized by their shape, diameter,
wattage (or length) and type of phosphor. The standard designations for
fluorescent lamps are shown in Figure 3-2. Refer to lamp manufacturer
catalogs for manufacturer-specific lamp designations.

Rapid-Start Linear Fluorescent Lamps

One of the more common fluorescent lamp types is the 40W, 1-1/
2-inch diameter 48-inch linear fluorescent lamp. The designation for this

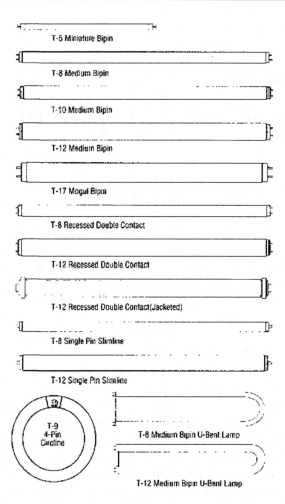

Figure 3-1. These are the most common full-size fluorescent lamp sizes and shapes. Although most full-size fluorescent lamps are manufactured as linear lamps, a variety of circular and U-shaped lamps are also produced. *Courtesy: Philips Lighting Company.*

lamp is **F40T12**, where the F stands for fluorescent, 40 stands for watts, T stands for tubular shape and 12 stands for the diameter in multiples of 1/8-inch. Similarly, the F32T8 lamp is a 32W linear (tubular) fluorescent lamp that is one inch (8 × 1/8 inches) in diameter.

The standard lamp designations can also describe the type of phos-

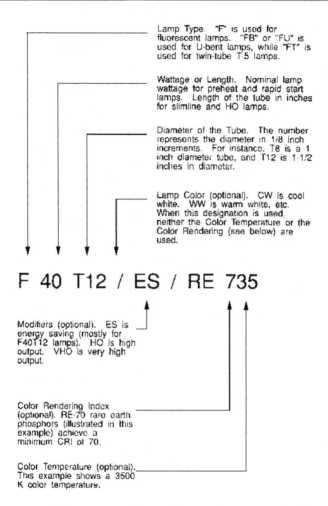

Lamp Type. "F" is used for fluorescent lamps. "FB" or "FU" is used for U-bent lamps, while "FT" is used for twin-tube T-5 lamps.

Wattage or Length. Nominal lamp wattage for preheat and rapid start lamps. Length of the tube in inches for slimline and HO lamps.

Diameter of the Tube. The number represents the diameter in 1/8 inch increments. For instance, T8 is a 1 inch diameter tube, and T12 is 1-1/2 inches in diameter.

Lamp Color (optional). CW is cool white. WW is warm white, etc. When this designation is used, neither the Color Temperature or the Color Rendering (see below) are used.

F 40 T12 / ES / RE 735

Modifiers (optional). ES is energy saving (mostly for F40T12 lamps). HO is high output. VHO is very high output.

Color Rendering Index (optional). RE-70 rare earth phosphors (illustrated in this example) achieve a minimum CRI of 70.

Color Temperature (optional). This example shows a 3500 K color temperature.

Figure 3-2. Use this as a road map for interpreting the designations used for identifying fluorescent lamps. *Courtesy: The California Energy Commission.*

phors used inside the lamp. For example, the linear 32W T8 lamp with a CRI rating of 75 and a color temperature of 4100K will have a designation of F32T8/741. The same 4100K lamp that uses a higher-grade triphosphor with a CRI of 85 will be described by the designation, F32T8/841.

Slimline, High-Output and
Very-High-Output Linear Fluorescent Lamps

The rules for designating lamps are different when the lamps are designed for use with slimline or high-output ballasts; these lamps are most common in lengths of 4-8 ft. Instead of the first number referring to wattage, it refers to the lamp's *length* in inches. For example, the 8 ft., 75W, 1-1/2-inch diameter slimline linear fluorescent lamp with a CRI of 75 and a color temperature of 3500K is designated as a F96T8/735 lamp, where the 96 refers to its 96-inch nominal length. Similarly, the 6 ft., 55W, 1-1/2-inch diameter slimline cool-white lamp is designated as F72T12/CW.

High-output (HO) and very-high-output (VHO) lamps—so named for their increased light output compared to the standard versions of the same-size lamps—require more amperage to start and to operate. As such, they require a special ballast designed for either HO (800mA) or VHO (1500mA) lamp operation. These lamps are designated with HO, 800, VHO, or 1500 along with the other information described above. For example, a high-output cool-white 8 ft. lamp would be listed as F96T12/CW/HO.

U-Shaped Fluorescent Lamps

Full-size lamps come in a variety of U-shaped configurations, typically to fit into 2×2 (2 ft. × 2 ft.) (or smaller) luminaires. Manufacturers have different ways of identifying their lamps, but the designation will include either a "B" for "Bent" or a "U" (for obvious reasons). For example, the Philips Lighting 31W T8 U-lamp with a CRI of 75 and a color temperature of 4100K is listed as FB31T8/TL741. (Note that TL is a Philips designation.)

T5 Twin-tube Fluorescent Lamps

Another type of fluorescent lamp is a variation of the twin-tube compact fluorescent lamp. T5 twin-tube lamps are normally classified as compact fluorescent lamps, but with lengths as long as 22-1/2 inches, they can be used in 2×2 fluorescent luminaires. Designations for these lamps are manufacturer-specific, such as DL (for Osram Sylvania), BX (for General Electric) and PL (for Philips).

T5 Linear Fluorescent Lamps

Continuing the trend toward smaller, more compact lighting systems for general lighting applications, major lamp manufacturers have introduced linear T5 lamps (with compatible electronic ballasts). These lamps are only available in *metric* lengths, which are shorter than the standard 2 ft., 3 ft. and 4 ft. lamp lengths used in North America. Although these systems provide high efficacy (slightly higher than T8 systems), their use is primarily in new fluorescent luminaires that are specifically designed to accommodate the unique dimensions of these lamps.

Energy Saver Lamps

Lamp manufacturers produce a series of reduced-wattage "energy-saver" fluorescent lamps. Each manufacturer uses its own designation for these types of lamps. For example, General Electric uses WM for *Watt-Miser*, Osram/Sylvania uses SS for *Super Saver*, and Philips uses EW for *Econ-o-Watt*. Further complicating matters, the designation for the 4 ft. energy saver lamps refers to the wattage of the lamp that it is designed to replace! For example, the General Electric, 4 ft. 34W 1-1/2-inch cool white energy saver fluorescent lamp is designated as F40T12/CW/WM, because it is designed to replace the 40W T12 lamp.

Performance Checklist for
Full-size Fluorescent Lamps

When selecting full-size fluorescent lamps for an upgrade, consider the factors that are described below.

Efficacy (Lumens Per Watt)

To achieve the highest lamp efficacy, specify *triphosphor* fluorescent lamps with a CRI of 82-85. However, it is difficult to judge the efficacy of a fluorescent lamp apart from the ballast to which it is connected. Refer to the system lumens and system wattage values published in manufacturer literature.

Lamp Lumen Depreciation

As a lamp ages through use, its light output gradually declines. This change is called lamp lumen depreciation (LLD) and is expressed as a percentage of "initial" lamp light output. Because wattage remains essentially the same over lamp life, LLD causes the lamp to lose efficacy as it ages. Triphosphor lamps—particularly those with a CRI in the 80s—tend to maintain the highest percentage of initial light output over their life.

Color Rendering Index

From Table 3-1, it is clear that the most energy-efficient fluorescent lamps also offer excellent color rendering performance. Most tri-phosphor lamps have a CRI rating of at least 75, which is considered excellent. Using high color rendering lamps will improve the aesthetics of the illuminated space and may even increase the occupants' perception of the light level.

Note, however, that the lamps with the *highest* color rendering performance must sacrifice efficacy to achieve near-perfect color rendering. Therefore, for maximum efficacy, choose fluorescent lamps with a CRI rating in the 80s.

Lamp Life

Most 8 ft. fluorescent lamps are rated to last at least 12,000 hours, and most 4 ft. lamps are rated to last 20,000 hours. Although lamp life (and its effects on lamp replacement and disposal costs) will affect lighting upgrade economics, it can represent only about 10 percent of the lamp's life-cycle cost, when considering the cost of energy used to operate the lamp over its life. Fluorescent lamp life is affected by several factors, most of which are controlled by the ballast. If the ballast displays the CBM (Certified Ballast Manufacturers) label, it meets ANSI (American National Standards Institute) standards for proper starting voltage, starting and operating cathode voltage, cathode preheat time, glow cur-

Table 3-1. Effects of lamp phosphor on 2-lamp F32T8 system performance.

CRI	2-Lamp System Wattage	System Lumens (Maintained)	System Efficacy (Maintained)
75	62	4,565	74 lm/W
85	62	4,992	81 lm/W
95	62	3,274	53 lm/W

Notes: System wattage based on 2-lamp electronic ballast with 0.88 ballast factor. Maintained system lumens account for lamp lumen depreciation at 40 percent of rated life.

rent, lamp current crest factor and ballast factor (lamp current). For more information about how the ballast affects lamp life, refer to the fluorescent ballast equipment overview later in this chapter.

Thermal Sensitivity of Fluorescent Lamps

The ambient temperature affects the light output and wattage used by fluorescent lamps. Note that when the ambient temperature around the lamp is significantly above or below room temperature, the performance of the system can change. Figure 3-4 shows this relationship for two common lamp-ballast systems: F40T12 lamps with a magnetic ballast and F32T8 lamps with an electronic ballast.

Note that the optimum performance for the T12 system occurs when the lamps are operated at room temperature (typically without any enclosure). However, the graphs show that the optimum operating temperature for the F32T8 lamp-ballast system is higher than for the F40T12 system. Because most fluorescent systems operate inside of luminaires

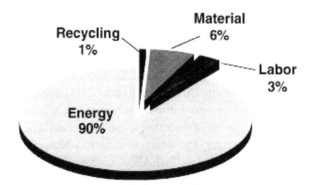

Assumptions: (2) T8 32W lamps
62W system wattage (w/electronic ballast)
Electricity at 7¢/kWh
Lamps at $2.65 each
Relamp labor at $1.50 each (group relamping)
Lamp life at 20,000 hrs
Lamp recycling at $0.50 each

Figure 3-3. Most of the life-cycle costs associated with the use of fluorescent lamps are in the energy they use. Therefore, economic decisions regarding fluorescent lamps are usually driven by energy efficiency, not lamp replacement costs. *Courtesy: EPA Green Lights.*

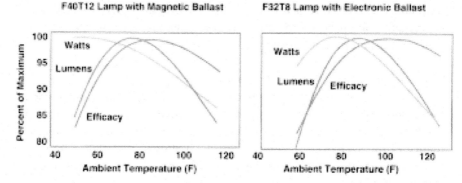

Figure 3-4. The performance of fluorescent lighting is significantly affected by the temperature inside the luminaire. *Courtesy: The California Energy Commission.*

where the temperature can be significantly higher than room temperature, T8 systems will operate closer to their optimal efficacy in most applications.

When reviewing product literature pertaining to fluorescent systems, performance data are based on testing under ANSI conditions—namely, bare tubes operating in 77°F air. To accurately estimate energy savings and lighting impacts, calculations should be based on thermally-corrected values for watts and lumens. When using uncorrected ANSI values for wattage, T12-to-T8 lighting upgrade calculations typically overestimate energy savings. The correction factors shown in Tables 3-2 and 3-3 may be used for correcting ANSI wattage and lumen values, respectively, based on operating temperatures. Note that *heat-removal* luminaires (see Figure 3-5) generally operate with improved performance over static luminaires.

For more detailed thermal performance data, refer to *Advanced Lighting Guidelines*, published by the California Energy Commission (CEC), the U.S. Department of Energy (DOE) and the Electric Power Research Institute (EPRI).

FULL-SIZE FLUORESCENT BALLASTS

All fluorescent lamps require an auxiliary piece of equipment called a ballast. Ballasts serve three main functions:

Table 3-2. Approximate ANSI thermal correction factors for *wattage*. Source: CEC, DOE and EPRI.

Lamp/Ballast System	3-Lamp Parabolic Louver	4-Lamp Prismatic Lens	3-Lamp Heat Removal	Industrial Strip
40W T12/magnetic	0.92	0.91	0.99	1.00
34W T12/magnetic	0.97	0.95	1.00	1.00
40W T12/electronic	0.94	0.92	0.99	1.00
34W T12/electronic	1.00	0.97	1.01	1.00
T8/magnetic	0.95	0.92	0.98	1.00
T8/electronic	0.94	0.90	0.99	1.00

Notes: Luminaires (except strip luminaires) are assumed to be recessed in a grid ceiling. Thermally corrected wattage = ANSI wattage x correction factor.

Table 3-3. Approximate ANSI thermal correction factors for *lumens*. Source: CEC, DOE and EPRI.

Lamp/Ballast System	3-Lamp Parabolic Louver	4-Lamp Prismatic Lens	3-Lamp Heat Removal	Industrial Strip
40W T12/magnetic	0.96	0.96	1.11	1.00
34W T12/magnetic	0.98	0.95	1.09	1.00
40W T12/electronic	0.97	0.97	1.09	1.00
34W T12/electronic	0.99	0.97	1.07	1.00
T8/magnetic	0.98	0.96	1.07	1.00
T8/electronic	0.98	0.95	1.08	1.00

Notes: Luminaires (except strip luminaires) are assumed to be recessed in a grid ceiling. Thermally corrected lumens = ANSI lumens x correction factor.

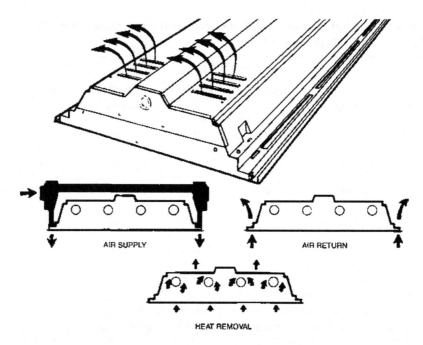

AIR SUPPLY

AIR RETURN

HEAT REMOVAL

Figure 3-5. Heat removal luminaires allow the lamps to operate at higher performance by reducing their temperature. In addition, air conditioning loads caused by the lighting system can be reduced with heat removal luminaires. *Courtesy: National Lighting Bureau.*

1.   The ballast provides the necessary starting voltage to strike an electric arc between the cathodes (which are located at each end of the lamp).

2.   Following lamp ignition, the ballast limits the lamp current, thereby regulating light output.

3.   During lamp operation, the ballast provides various power conditioning functions, including voltage regulation and power factor correction. Electronic ballasts also feature additional components for reducing harmonic distortion and electromagnetic interference.

Types of Fluorescent Ballasts
    The two general categories of fluorescent ballasts are magnetic and electronic. Within each category, there are several types of ballasts from which to choose.

Magnetic Ballasts

There are three types of full-size magnetic fluorescent ballasts that are currently used: standard "pre-1990" magnetic ballasts, "energy efficient" magnetic ballasts and hybrid (or cathode-disconnect) ballasts.

Standard "Pre-1990" Magnetic Ballasts. Standard "Pre-1990" magnetic ballasts are essentially core-and-coil transformers that are relatively inefficient in operating fluorescent lamps. The Federal Appliance Standard prohibits the issue of these magnetic ballasts.

"Energy-efficient" Magnetic Ballasts. The standard magnetic ballast described above was replaced by the new standard magnetic ballast—the "energy-efficient" magnetic ballast. Using high-quality copper wiring and enhanced ferromagnetic steel, energy-efficient magnetic ballasts use about 10 percent less energy to deliver the same light output provided by the pre-1990 ballasts. Note, however, that these "high efficiency" ballasts are the *least* efficient full-size fluorescent lamp ballasts available on the market today. More efficient ballast alternatives are described below.

Hybrid (or "Cathode Cut-out") Ballasts. Hybrid ballasts are energy-efficient core-and-coil magnetic ballasts with an electronic switch that disconnects the power to the cathode (filament) at each end of the lamp once the lamp starts. The cathode cut-out operation saves the wattage required to continually heat the cathodes, resulting in an additional savings of about 8 percent with T12 lamps and about 13 percent with T8 lamps. Hybrid ballasts are available in both partial-output (low-power) and full-output versions. For more information about the application of these ballasts in energy-efficient lighting upgrades, see Chapter 4.

Electronic Ballasts

In nearly every full-size fluorescent lighting application, electronic ballasts can be used in place of conventional magnetic core-and-coil ballasts. Electronic ballasts improve fluorescent system efficacy by converting the standard 60 Hertz (Hz) input frequency to a higher frequency, usually 25,000-40,000 Hz. Lamps operating at these higher frequencies produce about the same amount of light, while consuming 12-25 percent less power. Energy savings of up to 40 percent can be achieved by combining the efficacy of electronic ballasts with the efficacy of triphosphor fluorescent lamps. Other advantages of electronic ballasts include less audible noise, less weight, virtually no lamp flicker and the

Figure 3-6. Inside a magnetic (core and coil) fluorescent ballast. Courtesy: Advance Transformer Co.

Figure 3-7. Inside an electronic ballast. Courtesy: Advance Transformer Co.

ability to operate up to four lamps using a single ballast.

Specially-designed *dimmable* electronic ballasts are available that permit the light output of the lamps to be dimmed based on input from electronic dimming controls or from devices that sense daylight or occupancy. For specific lighting upgrade options with fixed-output electronic ballasts, see Chapter 4; for upgrades involving dimming controls and variable-output electronic ballasts, see Chapter 12.

Types of Eight-foot Fluorescent Systems

The longest lamps in the fluorescent family are the 8 ft. "F96" lamps. Typically used in retail and industrial applications, 8 ft. lamp-ballast systems are produced in three light-output categories as described below. Note that these three systems are available, but less popular, in shorter lamp lengths as well. Refer to Table 3-4 for a summary of 8 ft. fluorescent system characteristics.

Table 3-4. 8 ft. (F96) fluorescent systems.

System	Lamps	Ballasts	Efficacy (lm/W)[1]	Relative Light Output[2]	Lamp Lumen Depreciation[3]
Slimline	75W T12 60W T12 59W T8	Instant-start • Magnetic • Electronic	58-89 (83-117%)	100%	0.88
High Output (HO)	110W T12 95W T12 86W T8	Rapid-start • Magnetic • Hybrid • Electronic	53-81 (119-159%)	145%	0.87
Very High Output (VHO)	215W T12 185W T12	Rapid-start • Magnetic	42-44 (162-191%)	175%	0.75

[1]System efficacy is affected by lamp wattage, lamp CRI, ballast type and lamp lumen depreciation.

[2]System lumen output relative to a 2-lamp F96T12/CW magnetic system; full-output cool white values shown; ranges affected by lamp wattage, lamp CRI and ballast factor.

[3]Lamp lumen depreciation at 40% rated lamp life; assumes cool white lamps.

Slimline

The most common slimline systems consist of 2-lamp instant-start ballasts and compatible slimline lamps, providing approximately the same light output *per foot* as a two-lamp F40T12 (4 ft.) system. T12 slimline lamps are available in full-output 75W and reduced-output 60W versions for use with magnetic and electronic ballasts. In addition, 59W T8 slimline lamps are available for use with electronic ballasts. Common in all slimline lamps are the single-pin contacts used for instant starting.

High Output (HO)

High-output systems produce about 45 percent more light than slimline systems and require dedicated high-output (800mA) rapid-start ballasts. Like slimline systems, high-output lamps are available in standard and energy-saver T12 versions, and T8 HO systems have been introduced. HO lamps feature a recessed double-pin contact at each end of the lamp.

Very High Output (VHO)

The very-high output (1500mA) system is essentially obsolete because of its relatively low efficacy compared with slimline and HO systems. In addition, the VHO system's depreciation in light output is the most rapid of any lighting system. Similar to HO lamps, a recessed double-pin contact is used for transferring power from the socket to the VHO lamp.

Performance Checklist for Fluorescent Ballasts

The following factors should be considered when selecting a full-size fluorescent ballast for either retrofit or new installations. Most ballast catalogs will provide data in these categories to enable users to make informed upgrade decisions. In some cases, it is desirable to request independent laboratory testing data to confirm the manufacturer's performance claims. In addition, look for the CBM label that indicates that the ballast meets ANSI standards for starting voltage, starting and operating cathode voltage, cathode preheat time, glow current, lamp current crest factor and ballast factor.

Lamp-ballast System Efficacy (Lumens per Watt)

Select the combination of lamps and ballasts that is most efficient in converting electricity into light. For example, the 4-lamp *electronic* ballast used with 32W T8 lamps is one of the most efficacious fluorescent systems, yielding over 100 initial lm/W. By comparison, standard 40W lamps operating with "energy efficient" magnetic ballasts produce about 65 initial lm/W. Use the equation shown below to calculate system lm/W:

$$\text{Initial Lumens Per Watt} = \frac{\text{\# Lamps/Ballast} \times \text{Lumens/Lamp} \times \text{Ballast Factor}}{\text{Lamp \& Ballast System Wattage}}$$

These "initial" efficacy values should be corrected using the appropriate lamp lumen depreciation (LLD) factor to determine *maintained* efficacy values. Chapter 20 addresses lamp lumen depreciation and other light loss factors.

Ballast Factor (Light Output)
The light output from a lamp-ballast system is the product of lamp lumens and the ballast factor. The ballast factor is simply the percentage of the lamps' rated lumens that will be produced by the specified lamp-ballast combination. Although the ballast factor is a specific value for a given lamp-ballast combination, actual ballast factors can vary among ballast manufacturers and among lamp-ballast combinations. Table 3-5 lists the typical ranges of ballast factors for each ballast type.

Table 3-5. Typical ranges of ballast factors for full-size fluorescent ballasts.

Ballast Type	Range of Ballast Factors	
Magnetic Ballasts	0.93 - 0.95	with full-wattage lamps
	0.86 - 0.90	with energy-saver T12 lamps*
Hybrid Ballasts	0.81 - 0.95	
Electronic Ballasts	0.67 - 0.83	partial-output
	0.85 - 0.95	full-output
	1.05 - 1.30	extended-output

*Energy-saver lamps include 34W T12, 60W T12, 95W T12/HO and 185W T12/VHO lamps.

In applications where existing light levels are too high, *partial-output* electronic ballasts may be used. These ballasts operate fluorescent lamps with the high efficiency that is characteristic of electronic ballasts, but with specified reductions in both light output and energy consumption, resulting in further energy savings. In other applications, *extended-output* ballasts may be used to efficiently increase lamp lumen output beyond the lamps' rated values. Such applications include increasing light levels or recovering some of the light loss from delamping.

Lamp Flicker and Ballast Hum

Lamp flicker and ballast hum are annoyances commonly associated with fluorescent systems that operate at 60 Hz (on magnetic or hybrid ballasts). At this operating frequency, fluorescent lamps are turned off and on 120 times per second. Although lamp flicker may not be noticeable to some individuals, many complain about the distraction and discomfort that it can cause. Ballast hum can result from ballasts that vibrate at 60 Hz and are either improperly fastened to the luminaire or are nearing failure. Electronic ballasts solve these flicker and hum problems by operating fluorescent lamps at a much higher frequency. Operating at 25,000-40,000 Hz, electronic ballasts can eliminate complaints of fatigue and eyestrain frequently associated with flicker from conventional 60 Hz fluorescent systems.

Harmonic Distortion

Harmonic distortion is caused by a variety of electronic devices commonly found in modern office or industrial work places, including fax machines, computers, printers, photocopiers and variable-speed controls for mechanical equipment. Because these devices distort the smooth sinusoidal oscillation of current (amps) and voltage in electrical circuits, transient currents are produced, some of which contribute to increased amperage in the neutral conductor of four-wire, three-phase electrical systems, which could potentially overload the current-carrying capacity of these conductors should they be undersized. Total harmonic distortion (THD) is measured in milliamps and is sometimes expressed as a percentage of the undistorted current used by the lighting system. Because the harmonic currents are not used by the lighting system, they also contribute to reduced power factor. Table 3-6 shows the typical ranges of THD for each ballast type.

Because many utilities have not offered rebates on electronic bal-

Table 3-6. Total harmonic distortion (THD) ranges.

Ballast Type	THD Range
Energy-efficient and hybrid magnetic, 2-F40	15-20%
Energy-efficient magnetic, 2-F96	25-30%
Screw-in compact fluorescent electronic	18-80%
Industry-standard electronic	20% or less
Low-harmonic electronic	10% or less

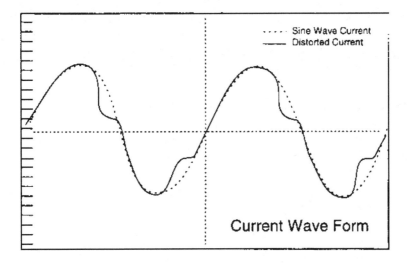

Current Wave Form

······ Sine Wave Current
———— Distorted Current

Figure 3-8. The use of electronic devices such as printers, fax machines, copiers, computers and lighting ballasts can distort the current and voltage waveforms in building electrical circuits. This distortion can interfere with the operation of sensitive electronic equipment and can add current to the neutral conductor in three-phase electrical circuits. *Courtesy: EPA Green Lights.*

lasts unless the THD is below 20 percent, nearly all electronic ballasts now meet this criterion. Some electronic ballasts with integrated circuits produce less than 10 percent THD. Because electronic ballasts use less current, maintaining the same *percent* THD will yield a *reduction* in the harmonic current. *Therefore, installing low-harmonic electronic ballasts can significantly reduce the total harmonic current on the trans-*

former circuit that serves the lighting system. Note, however, that reductions in harmonic distortion in a 277-volt (V) lighting system will not help to alleviate the harmonic currents in the 120V circuits that serve the THD-intensive fax machines, computers, printers and copiers. Operating at different voltages, the 277V circuits are electrically isolated from the 120V circuits.

Inrush Current

Inrush current is the current flow occurring at the instant the lighting circuit is switched on. Electronic ballasts that are designed to produce less than 10 percent THD may cause excessive inrush currents. High inrush currents can damage light switches, occupancy sensors and lighting control contactors (relays). In some cases, high inrush current can trip circuit breakers. However, ballasts with THD between 10 and 20 percent do not appear to be causing inrush current problems.

Lamp Compatibility

Not all lamps work with all ballasts. For example, T8 lamps (265mA) are designed to work with T8 (265mA) ballasts, and high-output T12 lamps (800mA) lamps are designed to work with high-output 800mA ballasts. Some electronic ballasts with integrated circuits can adapt to operate both T8 (265mA) and T12 (430mA) lamp types. Also, lamps with only one electrical contact at each end require operation with an instant-start ballast.

Minimum Starting Temperature

Nearly all fluorescent ballasts are designed to reliably start the lamps at a minimum ambient temperature of 50°F. However, due to the design of energy-saver T12 and T8 lamps, their minimum starting temperature is 60°F. Table 3-8 outlines the typical minimum starting temperatures. Refer to ballast manufacturer literature regarding minimum starting temperatures for a specific lamp-ballast combination.

T8 Instant-Start Versus T8 Rapid-Start

Although T8 *lamps* (up to 5 ft. in length) are classified as rapid-start lamps, electronic *ballasts* can be designed to start these lamps in either the rapid-start or instant-start mode. Instant-start ballasts provide improved system efficacy, but lamp life may be reduced if the system is switched more frequently than 12 hours per start. For more information

Table 3-7. Typical ballast inrush currents. *Source: Osram/Sylvania, Inc.*

Ballast Type	Typical Inrush Current Per Ballast
Magnetic Ballasts	5-7 amps
Electronic Ballasts (10-20% THD)	10-15 amps
Electronic Ballasts (<10% THD)	20-35 amps

Table 3-8. Minimum fluorescent system starting temperatures.

Ballast Type	Typical Minimum Starting Temperature
Magnetic	
with T8 lamps	+50°F
with standard T12 lamps	+50°F, 0°F
with all energy saver T12 & T8 lamps	+60°F
with HO or VHO T12 lamps	+50°F, 0°F, -20°F
Hybrid	
with standard T12 or T8 lamps	+50°F
with all energy saver T12 & T8 lamps	+60°F
Electronic	
with 2 ft., 3 ft., 4 ft. T8 lamps	+50°F, 0°F
with standard T12 lamps	+50°F, 0°F
with all energy saver T12 & T8 lamps	+60°F
with 8 ft. slimline T8 lamps	+50°F
with HO T12 lamps	+50°F, 0°F, -20°F

about this performance trade-off, refer to the application guidelines in Chapter 4 regarding the T8 lamp/ballast upgrade.

Parallel Versus Series Wiring

Parallel wiring is a feature available in several models of T8 and T12 electronic ballasts. *Parallel* circuits operate each lamp independently; if one lamp fails, the others continue to burn. Conversely, if one lamp burns out in a *series* circuit, all lamps in the ballast circuit go out. Nearly all instant-start ballasts operate lamps in parallel. Although most rapid-start ballasts are series type, the trend in electronic ballast manufacturing is to offer more rapid-start/parallel-wired products.

Number of Lamps Per Ballast

Although most magnetic and hybrid ballasts are designed to operate only two lamps, some *electronic* ballasts offer the advantage of operating up to four lamps. The use of 3- and 4-lamp ballasts instead of 2-lamp ballasts (where feasible) can yield savings in material, labor and energy costs, because fewer ballasts will be required, and because these ballast systems are more efficient. Table 3-9 shows the gains in efficacy that result from the use of multilamp rapid-start ballasts. In applications with 2-lamp luminaires, consider "tandem wiring" pairs of 2-lamp systems to share single 4-lamp ballasts. Check with ballast suppliers to determine their maximum wire length between lamps and ballast for reliable operation; the maximum recommended wire length is typically 12-15 ft.

Table 3-9. Increasing lamps per ballast improves rapid-start system efficiency.

System Description	System Lumens (Typical)	System Wattage (Typical)	System Efficacy (Lumens Per Watt)
2-lamp F32T8 rapid-start	4,565	62	74
3-lamp F32T8 rapid-start	6,847	90	76
4-lamp F32T8 rapid-start	9,025	116	78

Assumptions: Full-output electronic ballasts, lamp CRI = 75, maintained lumen output at 40% rated life.

Note: T8 instant-start system efficacy is virtually unaffected by the number of lamps per ballast, typically about 79 lm/W, maintained.

Electromagnetic Interference (EMI)

EMI can interfere with radio transmissions, sensitive electromagnetic instrumentation (such as electrocardiograms) and low voltage or powerline communication signals. EMI is produced by a wide variety of electronic equipment, including both magnetic and electronic fluorescent ballasts. However, all ballasts must meet the EMI emission limitations in Part 17 of the Federal Communications Commission. Nevertheless, there are rare situations where the EMI emitted from fluorescent ballasts can cause interference in wireless communication and detection systems. Electronic ballasts usually emit greater EMI than magnetic or hybrid ballasts, and not all electronic ballasts produce the same levels of EMI. Ballast manufacturers can provide EMI emissions data to assist in product selection.

Life-Cycle Cost

When evaluating the economics of converting to a new ballast type, consider all costs and savings that will occur during the life of the ballast. The variables that affect the life-cycle cost of a ballast include: material cost, installation labor cost, existing ballast disposal cost, ballast life, maintenance costs, annual hours of operation, system wattage, electricity rates, inflation rates, consulting fees and cost of capital (discount rate). Several lighting upgrade analysis tools are available for evaluating life-cycle costs (see Chapter 18).

FULL-SIZE FLUORESCENT LUMINAIRES

A full-size fluorescent luminaire—or light fixture—is the complete fluorescent lighting unit consisting of lamps, lamp sockets, ballasts, reflector material, shielding media (lens, diffuser or louver) and housing.

The main function of the luminaire is to distribute the produced light using reflective and shielding materials. Many lighting upgrade projects involve replacing one or more of these components to improve luminaire efficiency. Alternatively, consider replacing the entire luminaire with one that is designed to efficiently provide the appropriate quantity and quality of illumination.

Types of Full-size Fluorescent Luminaires

Specifiers identify full-size fluorescent luminaires by the attributes shown below:

- Nominal size and mounting method.
- Shielding media (lens, diffuser, louver).
- Distribution pattern (direct, indirect, or combination).

Nominal Size and Mounting Method

One of the most common full-size fluorescent luminaires is the 2×4 fluorescent "troffer." Troffers are fluorescent luminaires that are designed to be supported by a T-bar grid system. Other common ceiling grid dimensions include 1 × 4, 2 × 2, 4 × 4 and 2 × 8. In addition to lay-in troffers, other common mounting methods include surface-mount and suspended.

Shielding Media

Most indoor commercial fluorescent luminaires use either a lens or a louver to prevent direct viewing of the bare lamps. Light that is emitted in the so-called "glare zone" (angles above 60 degrees from the luminaire's vertical axis) can cause visual discomfort and reflections that reduce contrast on work surfaces or computer screens. Selections of shielding media will have an important impact on efficiency, visual comfort and luminaire spacing for uniform illumination. Table 3-10 shows the efficiency and visual comfort probability (VCP) for various fluorescent shielding materials. Note the tradeoff that often occurs between efficiency and visual comfort.

Table 3-10. Fluorescent luminaire shielding media.

Shielding Material (For 2×4 Troffers)	Luminaire Efficiency (%)	Visual Comfort Probability (VCP)
Clear Prismatic Lens	60-80	50-70
Low-Glare Clear Lens	60-80	75-85
Deep-Cell Parabolic Louver	50-75	75-95
Translucent Diffuser	40-60	40-50
Small-Cell Parabolic Louver	40-65	99

Lenses. Lensed luminaires include those with flat lenses as well as "wraparound" clear lenses. Although lenses made from clear ultraviolet-stabilized acrylic plastic deliver the most light output and uniformity of all shielding media, they provide less glare control than louvered luminaires. Specially designed low-glare flat lenses, however, have a special optic design that limits the amount of light produced in the high-angle "glare zone," making them suitable for spaces with computer screens. Lenses are generally less expensive than louvers.

Translucent Diffusers. This shielding media is essentially obsolete for use in modern commercial spaces. Translucent diffusers can provide unsatisfactory performance for *both* luminaire efficiency and visual comfort. Because the light is absorbed by the translucent material, luminaire efficiency can be as low as 40 percent. And because this shielding media diffuses light in all directions, computer users may find annoying reflections of translucent diffusers in their visual display terminals.

Louvers. Louvers provide superior glare control and high visual comfort compared to lens or diffuser systems. The most common application of louvers is to eliminate the luminaire glare reflected on computer screens. "Deep-cell" parabolic louvers—with 5-7 inch cell apertures and depths of 2-4 inches—provide a good balance between visual comfort and luminaire efficiency. Although *small-cell* ("paracube") parabolic louvers provide the highest level of visual comfort, they reduce luminaire efficiency to as low as 40 percent. For retrofit applications, both deep-cell and small-cell louvers are available for use with existing luminaires.

Strip Fixtures. Strip fluorescent luminaires are those without any shielding media. Although they provide the highest level of luminaire efficiency, they provide the lowest level of visual comfort.

Distribution Pattern

One of the primary functions of a fluorescent luminaire is to direct the light to where it is needed. The light distribution produced by a luminaire is characterized by the IESNA as shown in Figure 3-9.

The lighting distribution that is characteristic of a given fluorescent luminaire is described using candela distribution data provided by the luminaire manufacturer. The candela distribution is represented by a curve on a polar graph showing the relative luminous intensity 360° around the luminaire—looking at a cross-section of the luminaire. This information is useful because it shows how much light is emitted in each direction and the relative proportions of downlighting and uplighting (see Figure 3-10).

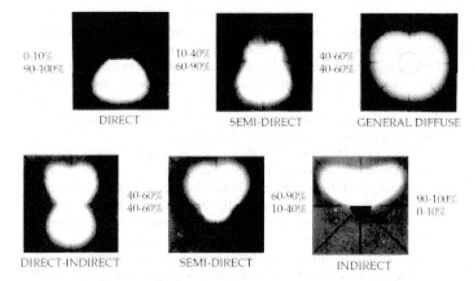

Figure 3-9. Luminaires are often identified by their type of light distribution as defined by the IESNA. *Courtesy: Illuminating Engineering Society of North America, New York City.*

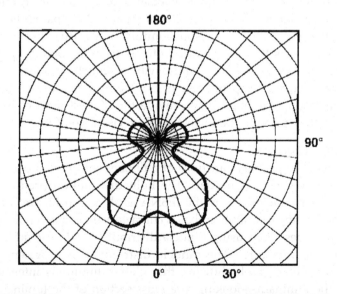

Figure 3-10. This candela distribution curve indicates that the luminaire provides downlighting with a small amount of uplighting. *Courtesy: EPA Green Lights.*

Performance Checklist For Fluorescent Luminaires

Fluorescent lighting upgrades should achieve specific goals with regard to luminaire efficiency, visual comfort and uniformity of illumination. However, these goals will be defined by the need for glare control in the spaces illuminated by full-size fluorescent lighting. For example, in a modern office space where computer screens are in use, higher goals for visual comfort may require a small sacrifice in luminaire efficiency and a decrease in luminaire spacing to achieve uniform illumination.

Luminaire Efficiency

The efficiency of a luminaire is the percentage of bare lamp lumens that actually exit the luminaire. This performance value is typically published on luminaire photometric reports. Generally, the most efficient luminaires have the poorest visual comfort (e.g., bare strip industrial luminaires); conversely, the luminaires that provide the highest visual comfort level can be the least efficient. Therefore, a lighting designer must determine the minimum efficiency and VCP needed for a space. The percentage of produced light that actually exits the luminaire can be increased by cleaning the luminaire, installing an improved reflector, or upgrading the lens or louver.

Coefficient of Utilization

The coefficient of utilization (CU) is the percentage of bare lamp lumens that reach the workplane. Although the efficiency of deep-cell parabolic luminaires can be less than lensed luminaires, the CU can be comparable for some deep-cell parabolic units with a "full chamber" design. Such a design features a curved or segmented reflector behind each lamp which integrates with the parabolic contour of the louver itself. Because most of the light is directed down to the workplane, rather than onto walls, the CU is improved. Although full-chamber designs can be used to minimize energy costs in new installations, the aesthetic concern about shadows on upper walls may need to be addressed with the use of accents, indirect lighting, or asymmetrical wall-washing louvers.

Visual Comfort

The visual comfort probability (VCP) is published in the luminaire's photometric report for direct illumination in a variety of room geometries. Spaces with computers should be illuminated with lighting systems that deliver a VCP of at least 80. Other commercial spaces without computers should achieve a VCP of at least 70.

Spacing Criteria

When new louvers or reflectors are installed to reduce high-angle glare, the light distribution becomes more narrow, and the uniformity of illumination can be compromised. Alternatively in new construction or renovation, high-VCP luminaires may require a closer spacing for uniform lighting, which will increase the number of luminaires. This problem becomes particularly acute in spaces with low ceilings. In such cases, select the most efficient luminaire that meets the minimum requirements for visual comfort, and space them within the criteria noted on the photometric report. For more guidance on specifying lighting upgrades in office applications, see Chapter 14.

Chapter 4

Full-size Fluorescent Upgrade Options

R ecent technological advances have created new opportunities for reducing energy consumption while enhancing the quality of fluorescent lighting systems. Today's fluorescent upgrade technologies can substantially improve source efficacy, luminaire efficiency, color rendering and visual comfort. In this chapter, we will review popular fluorescent lighting system upgrades and the strengths and weaknesses of the various strategies.

INTRODUCTION

Fluorescent upgrade investment decisions have been complicated by conflicting vendor claims and by the growing number of fluorescent upgrade options that have entered the marketplace. The material presented in this chapter sorts through the deluge of information that has bombarded many facility managers, and it provides clear direction for properly applying these new cost-saving options. In addition, troublesome application pitfalls can be avoided by noting the technology limitations in the "buyer beware" sections.

The fluorescent upgrades described in this chapter are presented in this order: 1) lamp upgrades, 2) ballast upgrades and 3) luminaire upgrades. These categories are introduced briefly below.

Fluorescent Lamp Upgrades

When addressing fluorescent lighting, saving energy does not mean sacrificing quality. Using triphosphor fluorescent lamps will help to maximize energy savings and improve color rendering and aesthetics. Although most fluorescent lamp upgrades are motivated by energy cost savings, some upgrades are necessitated by the prohibition of certain common lamp types. For example, the U.S. Energy Policy Act has prohibited the manufacture and importing of several types of lamps including the 40W (4 ft.) and 75W (8 ft.) "cool white" fluorescent lamps. So whether you are upgrading "because you have to" or because you are seeking to maximize lighting system efficiency and profit, consider the opportunities presented in this chapter.

Fluorescent Ballast Upgrades
Fluorescent upgrades should not be limited to simple lamp replacements. To realize the maximum profit potential in fluorescent lighting upgrades, consider upgrading the lamp-and-ballast *system*. That is, replace T12 fluorescent lamps *and* magnetic ballasts with the combination of triphosphor lamps and high-efficiency ballasts that will deliver the desired light level. With increasing demand for hybrid and electronic ballasts, there are now several options to consider when specifying fluorescent ballast upgrades.

Fluorescent Luminaire Upgrades
The upgrades in this chapter also address the light distribution and shielding functions of full-size fluorescent luminaires. The two-fold purpose of installing fluorescent luminaire upgrades is to improve luminaire efficiency (to save energy costs) and to improve visual comfort (to enhance worker productivity). However, these objectives must not be accomplished through compromises in recommended illumination levels or uniform illumination.

T8 LAMP/BALLAST UPGRADES

One of the most common upgrades for full-size fluorescent systems is the T8 lamp/ballast retrofit. Standard T12 lamps and ballasts can be replaced with new T8 lamps and compatible T8 ballasts to produce comparable light levels—while saving 30-40 percent in energy costs. Because T8 lamps operate at a reduced current (265mA for T8 versus 430mA for T12), they require a compatible ballast.

In addition to energy savings, T8 system upgrades can improve the color rendering performance of fluorescent lighting systems. T8 lamps are generally available in three versions of color rendering: A thin triphosphor coat results in a CRI rating of 75, a thick triphosphor coat produces a CRI rating of 85, and a special blend of triphosphors are used to produce a CRI rating of 95. T8 lamps usually *cost less* than T12 lamps with the same CRI rating.

T8 lamps and ballasts are available to replace almost every type of full-size fluorescent lighting system. Linear T8 systems are available in lamp lengths of 24, 36, 48, 60 and 96 inches. And 96-inch T8 lamps (8 ft.) are available in both standard-output and high-output (HO) versions. In addition, T8 lamps are produced in a variety of U-shaped lamps for use in 2×2 luminaires.

Application Guidelines

When using T8 lamps, specify lamps with a CRI rating of 85 to yield maximum efficacy and improved color rendering. Note that special T8 lamps with a CRI rating of 95 will sacrifice efficiency to achieve such unusually high color rendering. The triphosphor coatings not only improve color rendering and boost efficacy, they reduce lamp lumen depreciation over the lamp's life, resulting in further increases in overall system performance. The higher prices of the 85-CRI lamps can be justified by their improved efficacy performance as shown in Table 4-1.

Because T8 lamps have the same base pin configuration as T12 lamps, the T8 conversion does not require any modification to existing sockets. However, after converting to a T8 lamp/ballast system, it is possible to inadvertently replace a failed T8 lamp with a T12 lamp. Although a T12 lamp *may* be started and operated on a T8 ballast, lamp life and light output will be severely compromised. The best practice would be to convert all T12 systems to T8, and then *clear out the storeroom!* Keep only the compatible T8 lamps on hand.

Buyer Beware

With the exception of F96T8 lamps, all T8 lamps can be operated by either rapid-start or instant-start electronic ballasts. However, deciding whether to buy instant-start or rapid-start ballasts may not be an easy one. Instant-start ballasts can yield higher efficacy, but they can also reduce lamp life. Instead of "warming up the lamp" prior to ignition, instant-start ballasts deliver a much higher voltage "jolt" to the lamp cathodes, resulting in greater lamp wear per start. Figure 4-1 shows that the impact of ballast type on lamp life is determined by how often the lighting system is switched on and off.

Table 4-2 shows that the efficacy advantage of T8 instant-start ballasts becomes insignificant with 4-lamp ballasts. In many cases, the fi-

Table 4-1. Representative performance values: 4 ft. lamps (3-lamp systems).

Lamp Type	CRI	3-Lamp System Lumens[1]	System Wattage[2]	Lumens/ Watt	Relative Light Output
F40T12/CW (40W)	62	7,005	107	65	100% (base)
F40T12/CW/ES (34W)	62	6,087	92	66	87%
F40T12/835/ES (34W)	85	6,890	92	75	98%
F32T8/735 (32W)	75	6,847	90	76	98%
F32T8/835	85	7,488	90	83	107%

[1]Lumen ratings include effect of lamp lumen depreciation at 40% of rated lamp life.
[2]3-lamp system performance with rapid-start electronic ballasts (ballast factor = 0.88).

Table 4-2. System efficacy comparisons between F32T8 rapid-start and instant-start electronic ballasts. Assumes lamps with a CRI of 75. *Source: Manufacturer literature.*

Number of Lamps Per Ballast	T8 Rapid-Start (lm/W)	T8 Instant-Start (lm/W)
2 lamps	74	79
3 lamps	76	79
4 lamps	78	79

Note: Efficacy values include effect of lamp lumen depreciation at 40% of rated life.

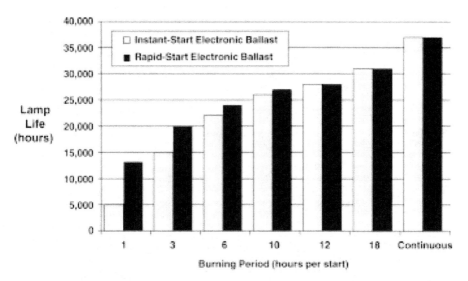

Figure 4-1. Compared to rapid-start ballasts, the use of instant-start T8 ballasts can reduce lamp life if the system is switched more frequently than 12 hours per start. Rapid-start T8 lamps may be an economical choice in applications where occupancy sensors are expected to frequently switch fluorescent systems. *Courtesy: EPA Green Lights.*

nancial advantage of using the more efficient instant-start ballasts offsets the costs associated with reduced lamp life. However, when occupancy sensors will be used and frequent switching is expected, consider using rapid-start ballasts to minimize reductions in lamp life.

T5 TWIN-TUBE LAMP/BALLAST UPGRADES

For maximum light output in 2×2 luminaires, specify T5 twin-tube fluorescent lamps and electronic ballasts. Although these lamps are normally listed as compact fluorescent lamps, the 22-1/2-inch twin-tube lamps can be used in many full-size fluorescent applications. T5 twin-tube lamps are also available in shorter lengths—as short as nine inches—and may be used in specially-designed luminaires.

Application Guidelines

Typically specified in new luminaires instead of retrofits, these compact systems may be used in higher ceilings or where high light

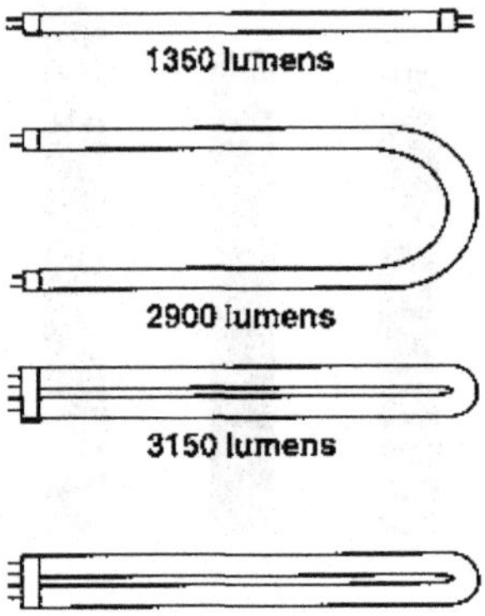

Lamp Type	Watts	Rated Life	System Efficacy*
F17T8 1350 lumens	17	20,000 hrs	76 lpw
FB31T8 2900 lumens	31	15,000 hrs	82 lpw
FT40T5 3150 lumens	40	20,000 hrs	74 lpw
FT50T5 4000 lumens	50	14,000 hrs	72 lpw

* 2-lamp/rapid start electronic ballast system

Figure 4-2. Facility managers and engineers have several choices of lighting systems for upgrading 2x2 luminaires. Selections are normally driven by light output requirements. *Courtesy: EPA Green Lights.*

levels are desired (such as in drug stores). In addition, these lamps are popular sources for use in indirect luminaires.

Buyer Beware

Because of the high surface brightness of these lamps, they should only be used in well-shielded fixtures. Unless the higher lumen output is required, consider using the more energy-efficient T8 lighting systems in 2x2 luminaires. Figure 4-2 compares the performance of several systems that may be used in 2x2 luminaires.

FULL-OUTPUT ELECTRONIC BALLASTS

Full-output electronic ballasts are high-frequency versions of conventional magnetic core-coil ballasts. Electronic ballasts operate fluorescent lamps more efficiently at frequencies greater than 20,000 Hz. The resulting increase in lamp efficacy, combined with reduced ballast losses, boosts fluorescent system efficacy by up to 30 percent. Other advantages

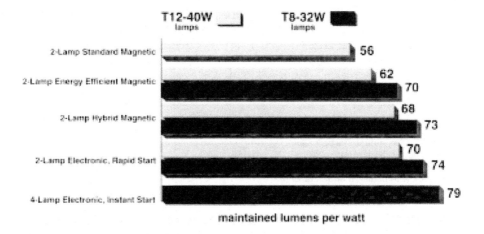

NOTE: *75 CRI lamps assumed*

Figure 4-3. T8 hybrid 2-lamp ballasts are nearly as efficacious as T8 rapid-start electronic 2-lamp ballasts. However, for maximum efficiency, choose T8 electronic ballasts that provide instant starting and/ or 4-lamp operation. *Courtesy: EPA Green Lights.*

are reduced weight, less humming noise, virtually no flicker and the capability to operate up to four lamps at a time.

"Full-output" electronic ballasts are rated with a ballast factor of at least 0.85, meaning that they must deliver at least 85 percent of the lamps' rated lumen output.

Application Guidelines

In nearly every fluorescent lighting system, full-output electronic ballasts can replace conventional ballasts, providing similar light output with significant reductions in energy consumption.

Although most magnetic ballasts are designed to operate only two lamps at a time, some electronic ballasts can simultaneously operate as many as four lamps. In applications with 2-lamp fixtures, consider "tandem wiring" pairs of two-lamp systems to share single 4-lamp ballasts, as shown in Figure 4-4. Check with ballast suppliers to determine the maximum wire length between lamp sockets and ballast for reliable operation. Note that tandem-wiring projects should be approved by the local building inspection authorities; in most cases, the use of pre-wired

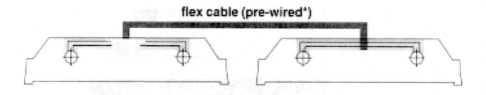

flex cable (pre-wired*)

* 3 wires for instant-start/parallel
  6 wires for rapid-start/parallel

Check with ballast manufacturer for
maximum tandem-wire distance

Figure 4-4. Tandem wiring minimizes ballast material costs when pairs of 2-lamp luminaires share a single 4-lamp ballast. Use parallel-wired ballasts in such applications so that in the event of a lamp failure, the remaining three lamps continue operating. *Courtesy: EPA Green Lights.*

flex cable is acceptable. The fixtures' UL listings are not affected by tandem-wiring.

Verify input wattage values for the proposed lamp-ballast combination because manufacturers' products will vary in this regard. Lower input wattages will increase energy savings and profitability, but will typically decrease light output.

Buyer Beware

Consider the performance criteria presented in Chapter 3 when shopping for electronic ballasts. These criteria include:

- System efficacy (lumens per watt).
- Light output (ballast factor).
- Harmonic distortion.
- Lamp compatibility.
- Minimum starting temperature.
- Instant-start vs. rapid-start.
- Inrush current.
- Parallel versus series wiring.
- Number of lamps per ballast.

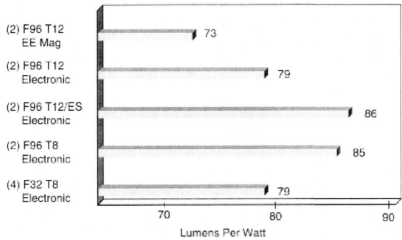

Figure 4-5. For maximum efficacy in 8 ft. slimline fluorescent systems, specify electronic ballasts with either energy-saver F96T12 lamps or F96T8 lamps. *Courtesy: EPA Green Lights.*

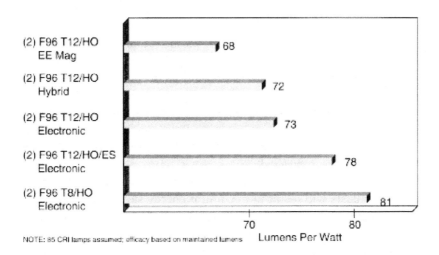

Figure 4-6. T8 high-output (HO) fluorescent lamps and compatible electronic ballasts provide highest efficacy of all 8 ft. HO fluorescent systems. *Courtesy: EPA Green Lights.*

- System interactions.
- Electromagnetic interference (EMI).
- Life-cycle cost.
- UL listing.
- Warranty.

PARTIAL-OUTPUT ELECTRONIC BALLASTS

Partial-output ("low-wattage") electronic ballasts operate fluorescent lamps at the same high efficacy as other electronic ballasts, but with specified reductions in both light output and energy consumption. These ballasts are produced with a ballast factor that is less than 0.85. (The ballast factor is the percentage of the lamps' rated lumens that will be produced by the specified lamp-ballast combination.) Partial-output ballasts typically have ballast factors in the range of 0.67 - 0.83; some switchable ballasts (with multiple ballast factors) feature even lower ballast factors when used on the "low" setting. Most electronic ballast brochures list the ballast factor for the various lamp-ballast combinations that are available.

Application Guidelines
Partial-output electronic ballasts should be used for minimizing electricity consumption where reduced illumination is acceptable. The availability of electronic ballasts with various output quantities enables specifiers to select ballasts with the appropriate output that will most closely meet the target light level. Because reduced-wattage electronic ballasts reduce energy consumption with little or no premium cost compared to standard-wattage electronic ballasts, *both* energy savings and profitability will be increased.

There are several applications where the use of reduced-wattage electronic ballasts will result in maximum energy savings and improved lighting quality:

Task/Ambient Lighting
By providing task lights at office workstations, the illumination required from the overhead lighting system is significantly reduced. In some cases, delamping alone will not reduce light levels to the 20-30 fc recommended for the ambient component of a task/ambient lighting

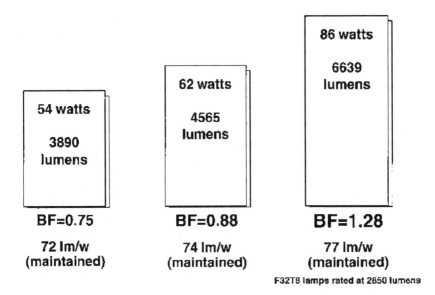

BF=0.75 | BF=0.88 | BF=1.28
72 lm/w (maintained) | 74 lm/w (maintained) | 77 lm/w (maintained)

F32T8 lamps rated at 2850 lumens

Figure 4-7. Major ballast manufacturers offer T8 fluorescent lamps in up to three categories of light output, each of which maintains relatively high system efficacy. Note that partial-output ballasts can maximize energy savings and profit where fixed reductions in illumination are desired. *Courtesy: EPA Green Lights.*

system. Reduced-output electronic ballasts can lower the light level and improve visual comfort through reduced luminaire brightness.

Alternative to Delamping

Particularly with parabolic louver fixtures, delamping can result in unfavorable fixture appearances. The use of reduced-wattage electronic ballasts enables all lamps on the ballast circuit to remain illuminated, thereby maintaining uniform brightness across the entire fixture while providing the appropriate amount of illumination on task surfaces.

Replacing 34W Fluorescent Systems

Conventional "energy-saver" 34W T12 lamps (which are reduced-output lamps) and magnetic ballasts can be replaced with 32W T8 lamps and partial-output electronic ballasts (with BF = 0.75-0.80) to achieve comparable light levels and save over 20 percent in energy costs. Table 4-3 illustrates this application.

Table 4-3. Partial-output ballast application—replacing 34W systems.

Lamps	Ballast Type (2-Lamp)	Ballast Factor	Maintained Lumens[1]	System Wattage
34W T12 / 62 CRI 2,280 design lumens[1]	Efficient Magnetic	0.87	3,967	72
32W T8 / 75 CRI 2,570 design lumens[1]	Partial-Output	0.77	3,958	56

[1]Design lamp lumens and maintained system lumens include effect of lamp lumen depreciation at 40% of rated lamp life.

New Fixture Layouts

Where ceiling heights are low and where low levels of illumination are specified, wider fixture spacing is needed to achieve the target illumination. In some cases, the required fixture spacing with full-output ballasts will be so great that non-uniform illuminance will result. Reduced-wattage ballasts can provide the target illuminance without exceeding the fixture's spacing criterion.

Buyer Beware

The same qualifications that apply to full-output electronic ballasts also apply to partial-output electronic ballasts. When specifying partial-output electronic ballasts, consider rapid-start ballasts which maintain cathode voltage during low-current operation, thereby preserving rated lamp life. For increased flexibility, consider installing continuously-dimmable or step-dimming (light-level switching) electronic ballasts (see next section).

DIMMABLE ELECTRONIC BALLASTS

Dimmable—or "controllable"—electronic ballasts are specifically designed to vary the light output of a fluorescent luminaire based on input from a light sensor, manual dimmer, occupancy sensor or scheduling system. Most dimmable ballasts are equipped with two additional low-voltage control leads that receive the signal directly from the controlling device. Other ballast designs receive the dimming signal over the line-voltage circuit. Although most dimming ballasts are available only in the 2-lamp configuration, 3-lamp and 4-lamp dimming ballasts have been introduced, which lower the material costs needed for dimming 3-lamp fixtures and 2-lamp fixtures that can be tandem-wired.

Application Guidelines

Daylight dimming is one of the most popular and cost-effective applications of dimmable electronic ballasts. Other applications include lumen maintenance control, manually-operated dimming and occupancy-sensed dimming. When more than one control device is used to control ballast output (such as a photosensor with an occupancy sensor), an integrated load controller is needed to determine the appropriate signal to send to the ballasts. For more information about dimming controls, see Chapter 12.

Buyer Beware

Do not let "sticker shock" stop serious consideration of this option. Although most controllable ballasts can cost twice as much as comparable fixed-output ballasts, new 3-lamp and 4-lamp products can yield acceptable returns on investment while providing maximum flexibility and energy savings. Refer to Chapter 18 for the methodology to follow for directly measuring the savings achieved by controllable electronic ballasts in a trial installation.

The controlling devices—photosensors, occupancy sensors, dimmers, etc.—must be compatible with the controllable electronic ballast. Check with the manufacturers to verify compatibility.

Harmonic distortion for most controllable electronic ballasts is very low due to the use of integrated circuit technology. Although the percentage of harmonic distortion does increase as the lamps are dimmed, the total harmonic distortion typically remains under 20 percent, even in low-current conditions.

Due to higher ballast losses, dimming electronic ballasts may draw 5-10 percent more energy at full light output than non-dimming electronic ballasts. A typical 2-lamp T8 dimming ballast may draw 64-65 watts at full output, compared to 58-62 watts for a non-dimming T8 electronic ballast.

When dimmed to 20 percent of full light output (maximum dimming for many controllable ballast designs), the system efficacy is about half of what it is at full output. Yet, this 80 percent reduction in light output is produced with about a 60 percent reduction in power (see Figure 4-8).

Lamp life may not be appreciably affected by dimming if the ballast is designed to maintain the proper cathode voltage when dimming.

When calculating energy cost savings expected from a dimming system, take into account the specific electric demand charge and rate structure; some rate schedules include a ratcheted demand charge that could diminish and/or delay cost savings resulting from reduced peak demand. Chapter 18 provides more detail on these factors.

LIGHT-LEVEL SWITCHING
ELECTRONIC BALLASTS

A low-cost method for providing occupants with a convenient choice of light levels is to install electronic ballast designs with "step-dimming" capabilities. These ballast designs allow users to select up to five different light levels from their wall switch. Another alternative is to

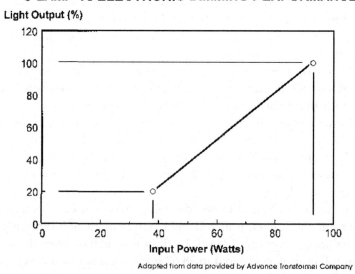

Figure 4-8. When controllable ballasts are dimmed to 20 percent of full light output, power input is reduced to about 40 percent of full wattage. Adapted from data provided by Advance Transformer Co.

install bi-level (or tri-level) switching electronic ballasts that preserve the dual-switching capability in most modern office spaces, while keeping all the lamps uniformly illuminated.

Application Guidelines

Where dual switching systems currently control 3-lamp fixtures, it may not be economical to replace both of the fixture's ballasts with electronic models to maintain the existing dual switching configuration. An alternative would be to tandem-wire four-lamp ballasts to operate the outboard lamps formally operated by the two-lamp ballasts, and a two-lamp electronic ballast could be tandem-wired to operate the inboard lamps—in pairs of luminaires. However, the added labor cost for tandem-wiring may exceed the added cost of installing one three-lamp light-level switching ballast per fixture.

Buyer Beware

Although step-dimming is an economical way to adjust light levels, occupants may prefer continuous dimming for establishing their preferred light level or for providing daylight-dimming control.

A low-cost alternative to the light-level switching ballast is the parallel-wired, fixed-output electronic ballast. The parallel wiring allows maintenance staff to lower light levels by selectively removing one or more of the lamps while the remaining lamps remain illuminated. Check with the ballast manufacturer regarding possible adverse effects resulting from operating the ballast without its full complement of lamps. In addition, determine if the appearance of partially delamped fixtures will be acceptable.

Verify that the step-dimming ballast meets ANSI standards for cathode voltage and operating lamp current to assure rated lamp life.

RETROFIT REFLECTORS WITH DELAMPING OPTION

Retrofitting luminaires with reflectors has been one of the more popular fluorescent upgrades. Yet, it remains one of the least understood retrofit options. Retrofit reflectors are devices that improve luminaire efficiency by reducing the amount of light absorption within the luminaire. Using more reflective materials behind the lamps results in a higher percentage of lamp light exiting the luminaire and reaching the task surfaces.

Applications

Retrofit reflectors may be the most cost-effective solution for restoring the performance of older, less-efficient luminaires. Common applications for fluorescent reflectors include commercial (shielded) fixtures and industrial strip fixtures.

Reflectors In Commercial Fixtures

Reflectors are commonly used for recovering some of the reduction in light output resulting from partially delamping 3-lamp or 4-lamp fluorescent troffers. Typically, the remaining two lamps in a 2×4 luminaire are relocated to positions centered on each side of the luminaire for maximum utilization of the reflector. The lamp relocation enhances light output and distribution, resulting in a more acceptable luminaire appearance. All ballasts and sockets used for operating the removed lamps should be disconnected in order to save additional energy and avoid confusing the maintenance staff.

Reflectors may be combined with the installation of higher-output lamps, higher output ballasts and/or improved lenses to minimize the light output reduction from delamping. To maintain the increase in luminaire efficiency that results from a retrofit reflector installation, reflector surfaces should be cleaned at regular intervals, following manufacturer's recommendations. In many cases, luminaire cleaning can have a greater effect on luminaire efficiency than installing a reflector.

Use reflectors only in situations where the additional lumen output is required to deliver the target light level. In task/ambient lighting designs, the suggested 20-30 fc ambient lighting target is typically achieved with delamping and converting to T8 lamps and electronic ballasts—without the use of a reflector.

Reflectors In Strip Fixtures

In addition to retrofitting enclosed luminaires, reflectors have been successfully installed in strip luminaires, typically used in high-activity retail applications. Such installations are designed to direct much of the wasted high-angle light down to the floor level. Because strip fluorescent luminaires are used for illuminating all surfaces including ceilings, walls and floors, it is important to limit the reflector's shielding angle so that walls and ceilings will continue to receive some illumination. Common energy-saving applications involve:

Without Reflector

With Reflector

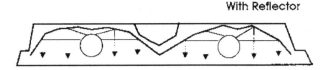

Figure 4-9. Retrofit reflectors are designed to minimize light absorption by limiting the number of light ray reflections inside the luminaire. *Courtesy: EPA Green Lights.*

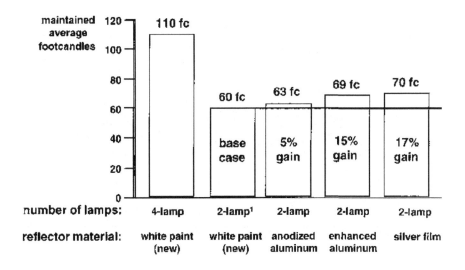

¹outboard lamps only

Figure 4-10. Based on independent tests, retrofit specular reflectors provided illuminance gains of 5-17 percent when compared to new fluorescent luminaires (with standard white enamel reflector surface). *Courtesy: EPA Green Lights, from* SPECIFIER REPORTS: SPECULAR REFLECTORS, *National Lighting Product Information Program, 1992.*

- Replacing 2-lamp VHO (1500mA) systems with 2-lamp HO (800mA) systems and specular reflectors.

- Replacing 2-lamp HO (800mA) systems with 2-lamp slimline fluorescent systems and specular reflectors.

- Replacing 2-lamp slimline systems with 1-lamp slimline systems and specular reflectors.

Buyer Beware

Are reflectors a wise investment for *your* application? Maybe. There are three factors that affect the performance of retrofit reflectors in improving luminaire efficiency and saving energy costs:

- The *material* (reflectivity) of the reflector.
- The *design* (shape) of the reflector.
- The *reflectivity* of the base luminaire's surface.

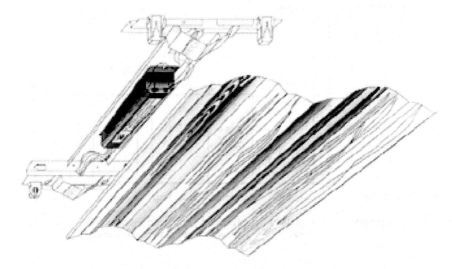

Figure 4-11. This assembly can be used to convert a 2x2 luminaire with two F40T12 U-lamps into one that uses two straight F17T8 lamps (with reflector option). Many of these kinds of assemblies are UL-listed because they are pre-wired and have raceways for socket wiring. The reflector in this diagram serves as the ballast housing. *Courtesy: Harris Lighting Systems.*

Reflector Material

Reflector materials are chosen based on a number of factors, including reflectivity, durability, cost and impacts on color shift. Table 4-4 summarizes these factors for each of the predominate reflector materials used in retrofit projects today. For maximum luminaire efficiency, select reflector materials with high *specular reflectivity*. Performing like a mirror, these materials can accurately control the light reflections, resulting in minimal light absorption within the luminaire. As an alternative to specular reflectors, *white reflector retrofits* are available that can improve an older fixture's efficiency while maintaining the original light distribution characteristics. These white reflectors can be significantly less expensive than specular reflectors.

Reflector Design

For maximum performance, the reflector should be designed to reflect light rays out of the luminaire using the fewest "bounces" possible—each bounce or reflection means that some of the light is being absorbed by the material. Using specular materials and ray tracing design methods, reflectors can be custom-designed to optimize luminaire light output. Although *custom* designs can deliver optimal performance, they are not always needed. Many "universal" designs are available in which the positions of the reflector and the lamp holders are designed to remain fixed in their relative positions, thereby holding the light distribution characteristics constant when used in a variety of luminaire types.

Reflectivity of the Base Luminaire

The most cost-effective applications of retrofit reflectors typically involve upgrading older luminaires that exhibit some degree of surface deterioration. Unlike today's high-reflectivity white powder-coat paint, the finish inside older luminaires may be dull, stained or corroded—a condition that simple cleaning cannot correct. In applications where it may be too expensive to replace older fixtures that have a dull or deteriorated finish, the installation of reflectors may be the most cost-effective method for restoring the fixture's efficiency.

Because of the wide degree of variation in the above factors, it is difficult to generalize about reflector performance in retrofit applications. For example, an anodized aluminum reflector installed in a new luminaire may only provide a 5 percent boost in light levels compared to a *new* 2-lamp luminaire without a reflector. Installing a higher performance material (either enhanced aluminum or silver film) using the same reflector shape may achieve about a 17 percent light output improvement compared to the new 2-lamp luminaire without a reflector. However, installing either reflector in an *older*, deteriorated 2-lamp luminaire may achieve a significantly *greater* relative improvement in luminaire output.

The best way to evaluate reflector performance is to install a trial upgrade in a space that is about 1,000 sq.ft. or more, and measure the performance using an illuminance meter. See Chapter 18 for a step-by-step procedure for objectively evaluating the performance of reflectors and other lighting upgrades in a trial installation. The key to this procedure is to start by cleaning and relamping the *existing* system before taking base-case light level readings.

There are alternatives to reflectors to consider. In some cases, the project economics will suggest that the reflector installation will not be cost-effective. Although the alternatives may not save as much energy, they may need to be considered when developing cost-effective retrofit projects. For example, consider other configurations that maintain the same number of lamps per fixture, but reduce light output, such as T8 lamps with partial-output electronic ballasts. If the existing lens or louver is absorbing a significant portion of the light, consider upgrading to a more efficient clear acrylic lens or a deep-cell parabolic louver. Also, when task lights are used in office applications, the ambient system may not require reflectors for providing the required 20-30 fc.

Table 4-4. Typical characteristics of reflector materials. Based on *Specifier Reports: Specular Reflectors*, **National Lighting Product Information Program, 1992.**

Reflector Material	Total Reflectivity	Specular Reflectivity	Material Cost	Durability
Silver Film	96%	94%	High	Low
Dielectric-Coated Aluminum	95%	92%	Highest	High
Anodized Aluminum	85%	80%	Medium	High
White Enamel	88%	3%	Low	Medium

LENS/LOUVER UPGRADES

Luminaire efficiency can be significantly improved by replacing inefficient or deteriorated shielding materials. Clear acrylic lenses provide maximum efficiency, and new "low-glare" *clear* lenses deliver this high efficiency with relatively high visual comfort. Many deep-cell parabolic louver designs also provide an excellent combination of efficiency and glare control.

Application Guidelines

The least efficient glare shielding materials—such as translucent diffusers or small-cell louvers—should be replaced with either clear acrylic lenses or deep-cell parabolic louvers. To determine impacts on visual comfort (glare control capability), refer to the product's VCP data, perform a trial installation, or visit an existing installation recommended by a lighting professional. Visual comfort is improved when light emitted at higher angles is shielded.

Figure 4-12. This retrofit deep-cell parabolic louver can replace the acrylic lens in most 2x4 lensed troffers. *Courtesy: A.L.P. Lighting & Ceiling Products, Inc.*

Deep-Cell Parabolic Louver Retrofits

Several manufacturers produce a series of deep-cell parabolic louvers for retrofitting lensed troffers. This retrofit is a relatively inexpensive way to improve visual comfort in modern office environments without significantly reducing luminaire efficiency. To perform this retrofit, first remove the existing lens and door frame. Next move the existing luminaire away from its original position in the ceiling grid, and place the new parabolic louver and door frame in the proposed ceiling grid location. The existing luminaire is then placed on top of the new parabolic louver door frame. When considering a retrofit deep-cell parabolic louver, make sure there is enough space in the plenum for the additional depth of the louver. And verify that the installation meets the support requirements of the applicable electrical and fire codes. Alternatively, small-cell parabolic louvers may be used if they can fit into the existing flat lens frame.

Low-Glare Clear Lenses

Specially designed clear lenses are an alternative to deep-cell parabolic louver retrofits. These unique lenses will reduce luminaire brightness at high viewing angles, resulting in VCP ratings in the 80s for most applications—suitable for offices with visual display terminals.

Standard Clear Lens Upgrades

In commercial spaces where computers are not in use, clear lenses are appropriate for glare shielding. They provide superior luminaire efficiency and room surface brightness. Over time, however, lenses can become yellow due to long-term exposure to low levels of UV light emitted from fluorescent lamps. Replacing the deteriorated lens with a UV-stabilized clear acrylic lens can be an inexpensive method for dramatically improving luminaire efficiency. This improvement in luminaire efficiency can be converted into energy savings if the increase in light output can be compensated by delamping or the use of partial-output ballasts.

Buyer Beware

Although retrofit small-cell parabolic louver panels (2-inch or smaller cells) are relatively easy-to-install retrofit products that provide high visual comfort (VCP > 0.90), they significantly reduce efficiency. Similarly, low glare *tinted* lenses also sacrifice efficiency in order to achieve high visual comfort.

When considering a deep-cell parabolic retrofit louver, ask the

Figure 4-13. The low-glare clear lens (on right) appears darker than the standard lens (on left) at normal viewing angles, thereby improving VCP. Yet, its efficiency is as high as clear acrylic prismatic lenses. *Courtesy: Holophane Corporation.*

manufacturer for VCP data and the candela distribution. When used in areas with computers, the louvers should have a candela distribution that meets the recommendations of the Illuminating Engineering Society's Recommended Practice No. 1 (IES RP-1) for glare control.

REPLACE WITH NEW DEEP-CELL
PARABOLIC LUMINAIRES

Instead of upgrading individual luminaire components, consider the labor savings and quality improvements that may be achieved by replacing existing luminaires with new luminaires that feature high-efficiency components such as T8 lamps, electronic ballasts and efficient louver shielding.

Deep-cell parabolic luminaires provide large-width louver cells (4-7 inches) to allow the light to efficiently exit the luminaire, while the

Safety Considerations

The UL (Underwriters Laboratories Inc.) Mark on a product means that samples of the product have been tested to nationally recognized safety standards and have been found to be reasonably free from foreseeable risk of fire, electric shock and related hazards. UL provides the third-party independent certification needed for consumers, inspection authorities, insurance interests and utilities.

Ballasts and luminaires are among the many products that bear the UL Mark. Some building inspection jurisdictions will require complete compliance with UL safety standards. Fluorescent luminaire upgrades, if specified and/or installed improperly can affect the integrity of the luminaire's UL Mark. Follow these guidelines to maintain the UL Mark on luminaires:

Ballasts, luminaires and other electrical devices that will be installed should be UL-LISTED. (Note that lamps are not UL-listed.)

Exit sign retrofit kits, retrofit reflectors (when used in place of ballast housings), socket relocation hardware and other devices consisting of parts or subassemblies intended for field installation in lighting fixtures should be UL-CLASSIFIED. For a manufacturer to obtain the UL-Classified Mark on their retrofit hardware, the manufacturer must have their device tested by UL in the specific type(s) of luminaires for which they are applicable. When the UL-Classified device is sold, it must be accompanied by documentation that indicates the fixtures/applications for which the device has been Classified, along with instructions for proper installation. The UL-Classified Mark must be affixed to the product and must include a UL control number.

To obtain a listing of manufacturers with UL-Classified products, call UL at (800) 704-4050 and purchase a copy of their "Electric Construction Materials Directory."

depth of the cells (3-4 inches) provides glare shielding for high visual comfort. The vertical surfaces of these louvers are parabolic in shape, thereby eliminating any light loss resulting from interreflection within the louver.

Application Guidelines

New deep-cell parabolic luminaires should be considered in offices where computers are used. Luminaires in these areas should provide shielding of high-angle light which can cause objectionable reflections in VDT screens, especially in large, open offices. The IESNA has published its recommended practice, RP-1, which addresses appropriate methods for lighting offices, including those containing computer VDTs. Before installing new luminaires, ask a lighting designer to verify the correct number and spacing of the luminaires based on published photometric data and the desired illumination level.

Buyer Beware

Although the *efficiency* of deep-cell parabolic luminaires is typically less than that of lensed fixtures, the *coefficient of utilization* for the highest performing deep-cell luminaires is comparable to that of standard lensed troffers. This advantage can be achieved by deep-cell luminaires that feature a *"full chamber"* design that aligns the parabolic louvers with a parabolic contour behind each lamp. This advanced design delivers more light down to the visual task, and less light is absorbed by the walls in the room.

To achieve their high coefficient of utilization and high visual comfort, deep-cell parabolic luminaires reduce light emissions at high angles, which can cause shadows to appear on the upper sections of walls. This aesthetic concern—known as the "cave effect"—can be addressed with

Figure 4-14. The small-cell parabolic louver can be an extremely inefficient solution to reducing glare from fluorescent luminaires. *Courtesy: Illuminating Engineering Society of North America, New York City.*

the use of accent lighting (e.g., wall sconces or wall-washers), indirect luminaires, or asymmetrical louvers.

REPLACE WITH NEW FLUORESCENT
UPLIGHTING LUMINAIRES

Uplighting luminaires direct the light upwards, reflecting off the ceiling to provide uniform, diffuse lighting on ceilings, walls and tasks. Uplighting can be accomplished with indirect fluorescent luminaires that direct at least 90 percent of the light toward the ceiling. Other systems—known as direct/indirect luminaires—provide a combination of uplighting and downlighting within the same luminaire.

Uplighting luminaires are usually suspended from the ceiling, although some luminaires are available that can be directly mounted on systems furniture partitions. With indirect systems, the light sources can be shielded from the view of occupants, which yields excellent visual

Figure 4-15. Indirect luminaires can enlarge the perceived size of the space by illuminating the ceilings and upper walls. *Courtesy: Peerless Lighting Corporation. Photograph by Richard Sexton.*

comfort. Compared with purely direct lighting systems (downlighting), the uplighting component can create the perception of a more spacious and pleasant environment, because ceilings and walls are illuminated.

Application Guidelines

Fluorescent uplighting is an excellent application for offices with computers. The indirect lighting can provide a uniform lighting distribution on the ceiling and walls, which helps to eliminate the distracting glare of light sources on display screens. Properly installed, uplighting luminaires meet the performance criteria of IES RP-1 for illuminating spaces with personal computers.

Another common application for uplighting is in partitioned spaces. Because the reflected light is more diffuse than light from direct systems, shadowing effects caused by the partition walls are reduced.

When installing uplighting luminaires, mount them according to the manufacturer's specifications. The correct suspension distance is critical for uplighting performance. If the sources are mounted too close to the ceiling, the resulting "hot-spots" will cause unwanted glare.

Because uplighting systems must be suspended below the ceiling, areas with low ceilings may be unacceptable. In such areas, consider installing uplighting systems that are specifically designed with a wide lighting distribution. In addition, furniture-mounted uplighting, or workspace-specific fixture suspension, may be preferred in low-ceiling areas.

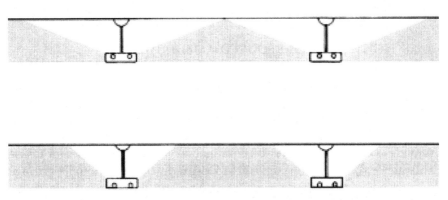

Figure 4-16. Indirect fluorescent luminaires should have a widespread light distribution to minimize distracting "hot spots" on the ceiling directly above the luminaire. *Courtesy: Illuminating Engineering Society of North America, New York City.*

Buyer Beware

Because uplighting luminaires are generally more expensive than direct systems, most uplighting systems are installed in new construction applications.

Although indirect systems provide relatively high visual comfort, some purely indirect systems have been described as "washed out" or "bland" without the contrast-enhancing qualities of direct lighting. Direct/indirect luminaires can address this concern. Alternatively, the combined use of purely downlighting luminaires with purely uplighting luminaires can create a pleasing aesthetic effect. The downlighting system can provide the needed ambient illumination in the interior area of a large space, while the uplighting system provides perimeter illumination and wall washing.

Uplighting systems yield a slightly lower workplane lumen efficacy (workplane lumens per system watt) than direct systems utilizing the same lamp-ballast combination. A highly reflective ceiling is essential for efficient uplighting systems. Workplane lumen efficacy will significantly decline when ceiling reflectances are below 80 percent. In addition, wall reflectance should be at least 50 percent.

Controlling luminaire dirt depreciation is of major importance for successful uplighting systems. Uplighting systems are more susceptible to dirt depreciation because dust will settle on the lens or interior surfaces. Regular cleaning is strongly recommended to minimize the effects of dirt depreciation.

TASK LIGHTING WITH REDUCED AMBIENT LIGHTING

Installing a task/ambient lighting upgrade can yield maximum energy savings in spaces illuminated with full-size fluorescent luminaires. To save energy in these applications, the light levels produced by the ambient (ceiling) lighting system can be significantly reduced to provide a comfortable level of illumination for short-term visual tasks and general circulation activity. To provide the additional lighting needed where visual tasks such as reading and writing are performed, energy-efficient fluorescent task lights may be used at individual workstations. Properly designed, this combined upgrade will yield significant energy savings while improving visual performance.

Application Guidelines

The task/ambient lighting approach can greatly simplify lighting upgrade designs. It is difficult to design a purely ambient lighting upgrade solution that meets the needs of *all* workers with respect to their specific visual tasks, preferences and visual capabilities. Task lighting can enhance user acceptance of the lighting system, because task lights can be adjusted to provide higher levels of illuminance where the user chooses. In situations where older workers require higher light levels, an additional (or higher output) task light could be provided.

Task/ambient lighting designs are best suited for office environments with significant VDT usage and/or where modular furniture can incorporate task lighting under shelves. For illuminating spaces with computer VDTs, the IESNA's RP-1 advises that the *ambient lighting* be designed to achieve *20-30 fc* on the work surfaces, with the *task lights* providing an additional *20-30 fc* to achieve the recommended maximum 50 fc needed for office-related visual tasks.

A variety of task light mounting methods are available: desk-clamp,

Figure 4-17. Compact fluorescent task lights can be repositioned by the user to minimize reflected glare and to illuminate specific task areas. *Courtesy: Dazor Manufacturing Corporation.*

free standing, under-shelf (with attachment hardware or magnetic mounting), systems-furniture panel-mount and wall-mounted using hinged brackets. Depending on the mounting method, the position of the compact fluorescent task light can be adjusted. Most compact fluorescent task lights can be repositioned horizontally; task lights with articulating arms or panel mounting hardware provide the added flexibility of vertical adjustments. The local task lighting should be installed below eye level so that it will not cause direct glare for the occupant.

Buyer Beware

Energy savings result when the energy saved from reducing the ambient lighting load exceeds the added energy used by the task lights. In some cases, the use of incandescent task lights may add more load than can be eliminated from the ambient lighting system. Compact fluorescent task lights provide high efficacy, low-glare workstation illumination.

Non-adjustable task light strips that are permanently mounted under cabinet shelves can cause reflected glare on work surfaces. To reduce reflected glare, specify compact fluorescent task lights that allow users to position the light to the side of the task. Alternatively, some linear undershelf task lights mask or shield the downlighting component to minimize reflected glare.

When installing portable task lights, consider the "plug loads" that will be added to the electrical distribution system. Be careful not to overload the amperage rating of the electrical circuits.

When considering a retrofit project to convert the existing general illumination system to a task/ambient system, perform a trial installation in a limited space to enable occupants to evaluate the system. Although employees may initially react with comments about the reduction in ambient light levels, most occupants will quickly adjust to the new lighting environment and will appreciate having control over their workstation illumination.

In new construction or renovation projects, most lighting designers will not normally specify task lights unless they are specifically requested. To ensure an energy-efficient, coordinated task/ambient lighting design, work with the designer and task lighting suppliers in selecting the appropriate design.

Chapter 5

Compact Lighting Equipment

T he original "compact" lighting system was the incandescent light bulb. However, the choices of light sources and luminaires in this category are growing. Applications previously reserved for standard incandescent lamps can now be served with tungsten-halogen, compact fluorescent, low-wattage HID and electrodeless lamps. In this chapter, we will discuss the characteristics of various types of compact light sources.

INCANDESCENT SOURCES

Although the incandescent light source still dominates the compact lighting market, other light sources are challenging its reign. In spite of these developments, however, there are many applications that will remain best served by incandescent technology.

Incandescent Lamp Types
There are many types of incandescent lamps, distinguished by the mixture of gases inside the bulb:

Standard Incandescents
Standard incandescent lamps that are rated at 40 watts or higher are filled with a mixture of inert gases. Nitrogen and argon are the most common. The fill gas helps to retard the rate of filament evaporation, thereby extending lamp life. Small lamps, those rated at less than 40 watts, typically contain a vacuum instead of a fill gas.

Halogen Incandescents

The halogen incandescent lamp—also known as tungsten-halogen or quartz-halogen—is another variation of the incandescent lamp. In a halogen lamp, the filament is contained inside a quartz capsule that contains a halogen gas. This small capsule causes the filament to operate at a higher temperature, which produces light at a higher efficacy and with a more "neutral-white" color than standard incandescents. The use of an optional infrared reflecting film on the inside of the halogen capsule can increase efficacy by reflecting waste heat back to the filament; the increased filament temperature yields a further improvement in lamp efficacy. The purpose of the halogen gas is to help preserve lamp life by capturing and redepositing the evaporated tungsten back on the filament. This tungsten-halogen process also keeps the capsule wall from blackening and reducing light output. Because the filament is relatively small, this source is often used where a highly focused beam is desired.

Incandescent Lamp Shapes

Incandescent lamps are produced in a wide variety of shapes. The shape is designated by one or two letters followed by a number. The numeric part of the lamp code denotes the maximum diameter of the lamp in 1/8-inch increments. For example, a lamp designated as an A19 is an arbitrary (A) shape and has a maximum diameter of 19/8ths of an inch (2-3/8 inches).

General Service

General service lamps include the typical household-variety incandescents, such as the very common A19 lamp and the more angular halogen bulbs (such as GE Lighting's TB19, Philips Lighting's T19, or Osram Sylvania's MB19).

Decorative

Incandescent lamps can serve in a vast number of decorative applications, including candle shapes (B, F) and globe shapes (G).

"R" Lamps

One-piece glass reflector ("R") lamps are available in two general categories: standard R lamps and energy-saver krypton R lamps. Because most R30 and R40 lamps have been discontinued because they don't

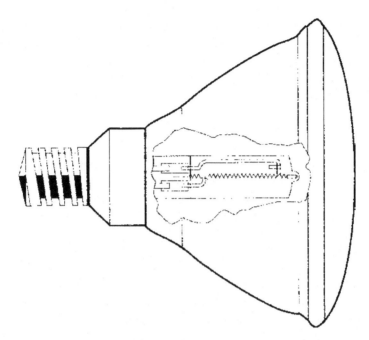

Figure 5-1. Most compact halogen lamps consist of a small quartz halogen capsule housed inside a rugged parabolic aluminized reflector (PAR) envelope. General-service "A-type" halogen lamps also contain a small halogen capsule inside the heavy glass enclosure. *Courtesy: California Energy Commission.*

meet the efficacy requirements of the U.S. Energy Policy Act (EPACT), a new category of lamps has emerged. The energy-saver R lamps include krypton that boosts the efficacy in order to comply with EPACT.

"PAR" Lamps

PAR lamps consist of a heavy glass cover and a parabolic aluminized reflector (PAR). This envelope is used for both standard and halogen incandescent lamp types. Compared with R lamps, the PAR construction provides better light beam control and durability in outdoor applications.

"MR" Lamps

MR (multimirror reflector) lamps are used in low-voltage applications where precise beam control is required for accent lighting. Typical sizes are MR16 and MR11 lamps. These small reflector lamps contain a tungsten-halogen capsule. Some MR lamps include a glass cover to filter ultraviolet light while others feature an open-reflector design. Similar to the MR lamps are the AR (aluminum reflector) lamps which provide a very consistent color temperature (chromaticity) over the lamp's life.

Incandescent Voltage Options

Line Voltage

Most incandescent lamps are designed to operate at the standard U.S. line voltage of 120 volts. However, selected long-life incandescents are available in alternative voltages of 130 and above. Refer to lamp manufacturer catalogs for details.

Low Voltage

Low-voltage incandescent lamps are used for accent and display lighting applications where tight beam control is necessary. Typically rated for 12V operation, these lamps require a transformer that converts incoming AC line voltage down to 12 volts DC. Accurate voltage transformation is required to maintain rated lamp life. The low-voltage operation enables lamp manufacturers to design more compact filaments, which enables the precise control of the light beam. Although low-voltage systems do not produce significantly more lumens per watt, they are capable of minimizing stray lumens so that the intended lighting target can be illuminated using fewer watts. Remember to include transformer energy use in efficacy evaluations.

Incandescent Lamp Bases/Sockets

There are several types of incandescent lamp bases that fit into compatible lamp sockets. By far the most common lamp base is the medium base. Other common bases include candelabra, intermediate and double-contact bayonet. Low-voltage halogen lamps feature unique 2-pin bases.

COMPACT FLUORESCENT SOURCES

Compact fluorescent lamps (CFLs) provide an energy-efficient alternative to incandescent lamps. When replacing incandescent light sources, compact fluorescent lamps can achieve energy savings of up to

75 percent. And the life expectancy of a compact fluorescent is about 10 times that of a standard incandescent lamp.

In recent years, much progress has been made in adapting compact fluorescent lamp technology to a growing number of incandescent applications. Hurdles regarding color quality, power quality, light projection and size have been cleared, opening up new avenues for saving energy in applications normally reserved for incandescent lamps.

Compact Fluorescent Lamp Shapes

One of the primary goals in compact fluorescent lamp design has been to reduce the overall size of the lamp to fit into conventional incandescent fixtures. Although the incandescent lamp is still smaller than most compact fluorescent lamps of equivalent light output, there are a variety of compact fluorescent lamp shapes that are designed to serve in specific incandescent applications:

Twin-Tube And Multiple-Tube Configurations

One of the earliest compact fluorescent lamp designs is the "twin-tube" lamp. This lamp essentially consists of two short T4 fluorescent tubes that are connected by a bridge, resembling a type of "U" lamp (with a very tight bend). To develop more compact designs, manufacturers also use multiple twin tubes in the same configuration, all connected by bridges to form a single light source. For example, major lamp manufacturers produce quad-tube (double twin-tube), triple-twin-tube, and even *quadruple* twin-tube designs. With multiple twin-tube lamps, the overall lamp length is shorter, but the efficacy may be slightly decreased, because the abundance of lamp material crowded into a small space tends to absorb some of the light before it leaves the lamp.

Another version of the twin-tube is the rapid-start T5 twin-tube (or U-tube). These lamps are designed for full-size fluorescent applications and are available in lamp lengths up to 22-1/2 inches. These lamps can be dimmed using compatible electronic ballasts.

Reflector

For downlighting and wall-washing applications, reflectorized compact fluorescent lamps provide a wide-angle beam of light. Due to the relatively large size of the compact fluorescent source (compared to an incandescent "point" source), it is difficult to produce a well-defined narrow beam using fluorescent technology. However, with improvements in reflector designs, manufacturers are now producing compact fluorescent "R-lamps" that produce candela distributions that are similar to the incandescent R40 flood lamps that they would replace.

Enclosed

Some compact fluorescent lamps include a globe enclosure that surrounds the light source. Many food service establishments use decorative globe-shaped compact fluorescents in pendant lamps mounted above the dining tables. Enclosed compact fluorescents should be considered whenever the lamp is not concealed by the luminaire and dimming is not required.

Compact Fluorescent Bases/Sockets
Screw-In

Compact fluorescent lamps with a standard medium screw-base offer an easy means to upgrade an incandescent luminaire. However, it is just as easy to screw in an inefficient incandescent lamp after the compact fluorescent lamp burns out. Therefore consider this "snap-back" effect when specifying compact fluorescent retrofits. Through proper maintenance training and lamp stocking practices, snap-back can be avoided.

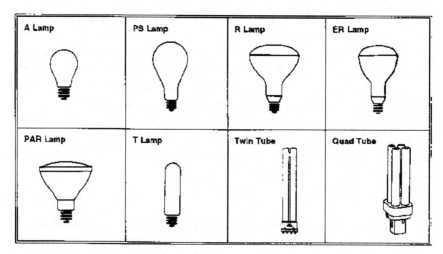

Figure 5-2. Common line-voltage incandescent and compact fluorescent lamp shapes. *Courtesy: EPA Green Lights.*

Hardwire Designs

To ensure that compact fluorescent upgrades will continue to deliver energy savings over the long term, consider hardwiring the new ballast and dedicated socket into the incandescent luminaire. The added labor cost for the installation may be worth the "insurance" of prolonged energy cost savings.

Compact Fluorescent Ballast Options

Buying a compact fluorescent lamp/ballast system involves several choices regarding the ballast.

Integral Versus Modular Units

There are two options to consider when specifying screw-in retrofit compact fluorescents. Integral units consist of a compact fluorescent lamp and ballast in self-contained units. When the lamp burns out, the entire package is discarded, including the still-functional ballast. The *modular* type of compact fluorescent retrofit consists of a dedicated compact fluorescent lamp (with a pin base) that can be separated from the screw-in ballast adapter (see Figure 5-6). Therefore, because a single compact fluorescent ballast may outlast four compact fluorescent lamps, modular designs offer savings in lamp replacement costs.

Magnetic Versus Electronic Ballasts

Although some magnetic ballasts are still used in compact fluorescent systems, the trend has been to utilize more electronic compact fluorescent ballasts. Electronic ballasts offer the advantages of 10-20 percent increased efficacy, reduced flicker, small size, instant starting, quiet operation and light weight. These advantages have helped improve user acceptance of compact fluorescent lamps.

Dimmable Compact Fluorescent Ballasts

Specialized compact fluorescent lamps and ballasts may be used to dim the light output of a compact fluorescent system. The compact fluorescent lamp must have *four* pins when used in a dimming application (the two additional pins are used for maintaining cathode temperature during the dimming process). The ballast must be a rapid-start dimming ballast, typically mounted in the fixture's housing frame due to its larger size.

Figure 5-3. These triple-U lamps provide instant-starting and high-efficiency illumination. *Courtesy: Philips Lighting Company.*

Figure 5-4. Enclosed compact fluorescent lamps are suitable for use in pendant fixtures and lanterns, where the lamps are directly visible by the occupants. *Courtesy: Panasonic Lighting.*

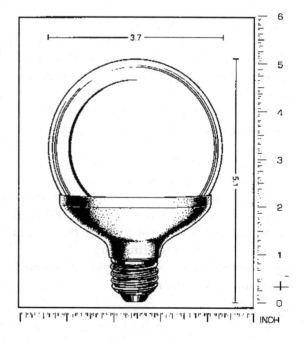

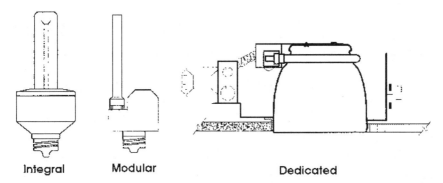

Integral   Modular   Dedicated

Figure 5-5 Choices for compact fluorescent upgrades include integral (one-piece) lamps, modular designs (2-piece) and dedicated (hardwired) retrofits. Modular and dedicated systems allow the longer-life ballasts to remain in the fixture as lamps are replaced. However, only the dedicated systems can assure long-term savings by eliminating "snap-back" potential. *Courtesy: California Energy Commission.*

COMPACT HID SOURCES

Low-wattage versions of high-intensity discharge (HID) lamps are finding cost-effective applications in commercial interior spaces. Metal halide and white high-pressure sodium lamps are the most popular compact HID sources. Most applications of compact HID systems are limited to new construction or renovation projects where new luminaires are specified.

Because these lamps feature a compact arc tube which closely resembles a point source of light, compact HID systems can be used in place of incandescent lamps where a point source is needed for light projection and "sparkle." And with the use of electronic HID ballasts, new compact-size HID luminaires can be used in spaces where older-technology HID systems were too bulky.

Like any HID system, these HID lamps require a dedicated ballast and are not easily dimmed. In addition, these systems require a prolonged warm-up time, and in the event of a momentary power interruption, a restrike period must elapse before the lamp can begin to warm up again. For more details about HID systems, refer to Chapter 7.

Metal Halide

With wattages below 35 watts, metal halide systems can meet many indoor lighting design requirements. New compact metal halide lamps with ceramic arc tubes are available to serve in place of incandescents with a color temperature of 3000-3500K and a color rendering index of 80-85. The efficacies of these low-wattage systems (less than 150 watts) are similar to compact fluorescent. Spot and flood lighting can be provided using either R or PAR envelopes that contain the metal halide lamp. Metal halide sources as small as the PAR20 narrow spot lamp are now available, which can produce incandescent-like color and sparkle.

White High-pressure Sodium

A variation of high-pressure sodium systems, white high-pressure sodium lamps and specially-designed electronic ballasts work together to efficiently deliver incandescent-like point-source lighting. Available in wattages ranging from 35 to 100, the color temperature is very close to traditional incandescent (2600-2800K) and the color rendering index can exceed 80. Although the efficacy of the white high-pressure sodium system is less than compact fluorescent (and is actually comparable to mercury vapor), the white high-pressure sodium system is the high-efficiency option that most closely matches the lighting characteristics of the incandescent lamp. Note that each manufacturer's white sodium lamp requires a unique electronic ballast for their specific lamp; ballasts from other manufacturers may be incompatible.

PERFORMANCE CHECKLIST

There are several physical and operating characteristics of energy-efficient compact sources that limit their universal application as incandescent replacements. Follow the guidelines below when selecting compact source options for incandescent applications. Many of the performance variables are summarized in Table 5-3.

Workplane Lumen Efficacy

Although choosing light sources with high efficacy is important, it is more important to specify lighting systems that efficiently produce and deliver the light to the intended target using the least watts. In high-ceiling applications for example, there are situations where reflectorized com-

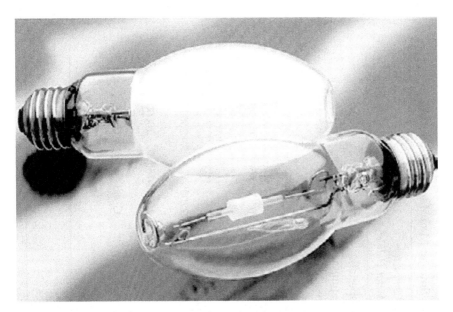

Figure 5-6. New low-wattage metal halide lamps with ceramic arc tubes replace incandescent lighting in non-dimming applications. Used with a specially designed electronic ballast, the system provides high efficacy and excellent color rendering. The small size of the arc tube allows for tight beam control in directional lighting applications. *Courtesy: Philips Lighting Company.*

pact fluorescent lamps will produce the highest *source efficacy* among the alternatives, but their diffuse light output can yield a lower *workplane lumen efficacy* compared with directional compact halogen lamps. In these situations, the most "efficient" lighting system depends on the specific room geometry or the lighting effect chosen for the application. The example in Table 5-4 illustrates how a compact halogen source may provide the target illumination using fewer watts than compact fluorescent.

The workplane lumen efficacy value (in workplane lumens per watt) is the average footcandles on the workplane (or display) divided by the unit power density (watts per sq.ft.) of the lighting system. The average footcandles can be determined from a trial installation or from lighting design calculations. In display lighting applications where precise beam control is needed, workplane (or display) lumen efficacy is often maximized with halogen sources.

Table 5-3. Characteristics of compact light sources. Source: Manufacturer literature.

	Compact Halogen	Compact Fluorescent	Compact Metal Halide[2]	Deluxe HPS[2]	White HPS
Dimming	standard incandescent dimming	special equipment needed	special equipment needed	special equipment needed	special equipment needed
Maintained Efficacy (lm/W)[1]	18-33	24-68	38-60	42-55	22-27
Lumen Maintenance	0.99	0.80-0.90	0.75-0.85	0.85-0.95	0.85-0.95
CRI Rating	100	82-86	60-90	65	70-85
Color Temperature (K)	2800K-3100K	3000K-4100K	3000K-3500K	2200K	2600K-2800K
Rated Lamp Life (hrs)	2000-5000	9000-12000	5000-15000	10000-15000	10000
Beam Control	Excellent	Poor	Good	Good	Good
Hot Restrike Time (min.)	0	0	5-15	1-2	1-2

[1]Includes effects of lamp lumen depreciation and ballast losses.
[2]Wattages less than 150.

Lamp Lumen Depreciation

For relatively short-lived incandescent and halogen sources, the effects of lamp lumen depreciation are relatively minor—particularly in halogen sources where the halogen cycle keeps the quartz bulb clean from the by-products of burning tungsten. However, because compact fluorescent lamps can operate for more than 10,000 hours over their life, the effects of lamp lumen depreciation in these systems can be more severe. For example, compact fluorescent lamps typically deliver an average of 85 percent of their initial lumens over their rated life. This depreciation effect means that typically one watt of compact fluorescent can replace only about 3-4 watts of incandescent lighting, instead of the 4-5 watts normally advertised in compact fluorescent product literature. In addition, lamp lumen depreciation effects should be considered for other long-life compact sources such as metal halide and white sodium systems.

Color Rendering Index

For maximum color rendering, incandescent and halogen sources are the clear favorite, with a CRI of 100. But do not overlook compact fluorescent lamps for use in high CRI applications; the typical range of CRIs for compact fluorescent lamps is 82-86, which is considered excellent. Unless the visual task requires a high degree of color discrimination (as in a color matching task), 82-CRI lamps provide the high-quality color rendering properties that occupants appreciate.

Lamp Life

Both the compact fluorescent and HID options provide lamp-life advantages over standard incandescent lamps. The long life of compact fluorescent or HID lamps should be considered in life-cycle cost analyses to justify the higher cost of these lamps. In commercial applications, the savings in lamp replacement labor due to the long life of these lamps can contribute to the overall profit generated by a compact fluorescent upgrade.

Thermal Sensitivity

Unlike incandescent, halogen or HID lamps, compact fluorescent lamps are sensitive to temperature for starting, efficacy and light output. Although their rated performance is based on operating in an ambient temperature of 77°F, many compact fluorescent lamps operate outdoors in much colder environments, as well as in unvented incandescent luminaires where the temperature is much higher than 77°F. In both cases, the

Table 5-4. Downlighting performance comparison: Compact fluorescent versus halogen in high-ceiling example.

	Compact Fluorescent	Halogen
Number of Luminaires	16	4
Lamp Type	13W Vertical Quad Tube	50W Halogen IR Flood
Luminaire Type	Specular Downlight	Specular Downlight
Luminaire Wattage	15	50
Maintained Lamp Lumens	800	1000
Maintained Lamp Efficacy	53 lm/W	20 lm/W
Total Installed Wattage	240	200
Coefficient of Utilization	0.31	0.84
Maintained Footcandles	14.8	14.8
Workplane Lumen Efficacy	13.8 workplane lm/W	16.6 workplane lm/W

Assumptions: Room cavity ratio = 10.0; room size: 15 ft. x 15 ft.; ceiling height: 17-1/2 ft.; task height 2-1/2 ft.; light-colored walls and ceiling; medium-colored floors.

result is reduced light output and reduced efficacy.

The minimum starting temperature of the compact fluorescent lamp is governed by the ballast. Specify ballasts with a minimum 0°F starting temperature for most outdoor applications of compact fluorescents.

The lamp's sensitivity to temperature is a function of the lamp's chemistry. Some compact fluorescent lamps use a mercury amalgam instead of pure mercury in the process of generating fluorescent light. The amalgam enables the lamp to operate within 85 percent of peak efficacy over a very wide range of operating temperatures.

Life-Cycle Cost

Buyers of compact lighting systems must resist the temptation to look only at first cost when evaluating options. Although the incandescent lamp may carry the lowest price, it may in fact be the most expensive light source after considering its high energy cost and short life.

APPLICATIONS OF COMPACT SOURCES

Each of the compact sources described in this chapter has unique operating characteristics that qualifies it for specific applications. The following incandescent luminaires are targets for energy-efficient lighting upgrades, which can achieve rapid financial returns and improved lighting quality.

- Downlights.
- Track lights.
- Wall washers.
- Task lights.
- Pendant lamps.
- Surface luminaires.
- Table lamps.
- Outdoor luminaires.

Downlights
Compact Fluorescent Sources

Compact fluorescent lamps are preferred for many applications because of their low wattage, which can translate to low life-cycle costs. Compact fluorescent lamp retrofits should be considered for downlighting applications in ceiling heights up to 20 ft. Although retrofit dimmable compact fluorescents are becoming available, other technologies should also be considered for dimming applications, such as halogen-infrared PAR lamps. Trial installations are suggested in high-ceiling applications to determine the least-wattage solution that produces the target light levels.

The selection of compact fluorescents in downlighting retrofits should be influenced by the resultant light distribution characteristics. Too often, retrofit compact fluorescent lamps are simply screwed into downlight sockets without regard for the possible side-effects of direct and reflected glare. Because of the generally longer length of compact fluorescent lamps, they can extend into the occupant's line of view as well as cause distracting reflected glare on specular downlight surfaces.

This glare may be particularly annoying, because downlights are specifically designed for maximum visual comfort.

To minimize visual discomfort with compact fluorescent downlight retrofits, consider these options:

1. Install a reflectorized compact fluorescent with a relatively short overall length and a relatively high center beam candlepower. The reflector should be designed to direct the light downward, confining most of the light to a beam angle of less than 30 degrees.

2. Install a compact fluorescent downlight conversion kit. Most kits use the existing downlight's mounting frame and junction box, to which a new lamp socket housing, reflector and ballast are installed. Both the lamp socket and reflector are specifically designed for the specified compact fluorescent lamp. Retrofits generally take less than 10 minutes per fixture, because most parts clip or screw

Figure 5-7. New modular designs of compact fluorescent reflector lamps feature shorter overall lengths. The result is improved visual comfort in downlight retrofit applications. *Courtesy: Lumatech Corporation.*

in. Both one- and two-lamp retrofit kits are available. For maximum downlight efficiency, use kits that position the lamp horizontally or diagonally. Several downlight manufacturers now offer these retrofit kits for upgrading their standard incandescent downlights. The resulting installation is resistant to snap-back, while providing relatively high luminaire efficiency and visual comfort.

Halogen PAR Lamps

Halogen PAR lamps should generally be used in applications where the limitations of compact fluorescent lamps would discourage their use. Halogen PAR lamps are fully dimmable (as many downlighting systems are), they project well from high ceilings, they are very compact, they provide the highest visual comfort, and they are immune to performance sacrifices in relatively warm, unvented incandescent luminaires. For maximum efficacy, consider the *halogen infrared* PAR lamps. When specifying halogen PAR lamps for downlighting, choose the correct beam spread: too narrow of a beam spread (spot lights) will provide "hot spots" under each downlight, creating a distracting pattern that reduces the overall aesthetics of the space; too wide of a beam distribution will lower the downlight's luminaire efficiency as wide-angle light becomes trapped inside the can.

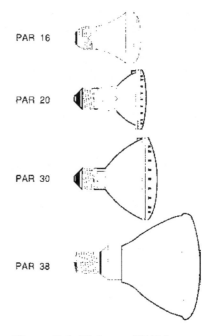

General Service (Non-Reflector) Sources

Many commercial-grade downlights are manufactured with a built-in specular reflector. These down-lights are designed to produce effective downlighting efficiency with the use of the common A19 or A21 incandescent bulb. When retrofitting these downlights, it is not nec-

Figure 5-8. Halogen PAR lamps are available in a variety of diameters for use in down-lighting and display lighting applications. *Courtesy: California Energy Commission.*

essary (and it can be physically impossible) to install upgrade lamps with integral reflectors. For economical retrofits in non-dimming downlights with integral reflectors and vertically oriented A-lamps, install compact fluorescent retrofits without integral reflectors. Again, verify that the length of the retrofit lamp does not contribute to unacceptable levels of direct or reflected glare. For economical retrofits in specular downlights with integral reflectors and *horizontally* oriented A-lamps, simply install a retrofit halogen A-lamp.

Track Lights

Track lights are typically used for creating high brightness levels on merchandise or art work, relative to their surroundings. For example, in retail lighting, feature displays should be illuminated to about five times the background illuminance. To limit stray light from contributing to background illuminance, tight beam control is essential. For this reason, halogen lamps are typically used for retail lighting; low-voltage halogen systems are preferred where the highest degree of optical control is desired.

Other applications of track lighting include wall washing and general ambient lighting. Refer to the section below regarding the use of compact sources in wall washing applications.

Wall Washers

Wall washing is a technique for brightening wall surfaces in an effort to balance brightness levels within a room, as well as to enlarge the perceived size of the space. In these applications, the "softness" of compact fluorescent reflector lamps or EPACT-compliant R-lamps can provide uniform wall washing when installed in track lighting equipment. (Because of the optical precision of halogen lamps, shadows and light intensity variations can detract from the desired wall-washing effect.) The use of wall-washing compact fluorescent track lights and sconces can also be used to add a decorative touch to an otherwise plain lighting design.

Task Lights

Traditionally, incandescent and linear fluorescent sources have been used in task lighting applications. However, as *compact* fluorescent sources have been introduced into modern task light designs, users now have greater control over the amount and location of their visual task illumination. Compact fluorescent task lights enable the source to be positioned to the side of the visual task in order to minimize reflected glare on glossy printed materials. Consuming far fewer watts than full-size linear (undershelf) systems, compact fluorescent task lights can be used in a task-ambient lighting upgrade to maximize energy savings in office lighting systems.

In some applications, users may prefer to use decorative task lights with compact halogen lamps. Although some of these task lights use halogen capsules without a glass cover for UV protection, most of these products do.

Pendant Lamps

Pendant (suspended) lamps are frequently used in restaurants for decorative illumination over dining tables. Most of these luminaires are specifically designed to accommodate incandescent A-lamps for illumination. However, "globe-shaped" compact fluorescent lamps are ideal

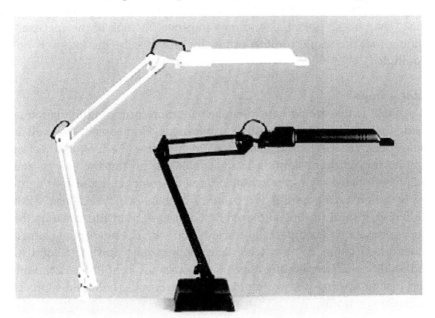

Figure 5-9. Compact fluorescent task lights give occupants control over their illumination. These task lights feature articulating arms that enable the user to position the task light for optimal task visibility. *Courtesy: Dazor Manufacturing Corporation.*

for these applications because they can maintain the decorative look of the luminaire, while providing energy-efficient, long-lasting, high quality lighting. (Note, however, that these globe-shaped fluorescent lamps should not be used on dimming circuits.) In applications where very large globes are used, non-dimming retrofit compact fluorescents could be used, but due to the potential for a high-temperature operating environment, *amalgam* compact fluorescent lamps should be specified for maximum efficacy.

Surface Luminaires

Where incandescents have been used in surface-mounted ceiling luminaires, consider installing new luminaires or retrofit systems that utilize either circline or 2-D lamps. In some cases, the physical constraints will require that the ballast be hardwired within the existing luminaire. Retrofit kits are available that convert conventional downlights to a translucent surface "round" luminaire using compact fluorescent lamps and electronic ballasts. This retrofit will increase luminaire efficiency, reduce "snap-back" potential, and may improve the look of the space, assuming that the visual comfort originally provided by the fixture is not required.

Table Lamps

Table lamps are perhaps the most common form of illumination in residential and hospitality lighting applications. Homeowners, as well as innkeepers, are most concerned about maintaining the look and feel of familiar incandescent lighting in their table lamps. But the energy and maintenance cost savings potential offered by compact fluorescent lighting is too great to pass up! The primary barriers of physical size, color rendering, instant starting and even first cost have been addressed by the latest generation of compact fluorescent products. The new spiral-shaped and triple-U lamps fit most table lamp applications, and the 82-86 CRI and 3000K color temperature appears fully incandescent to the occupant. And instant starting is now provided by electronically-ballasted compact fluorescent retrofits.

In addition to color quality and instant starting, another desirable characteristic of compact fluorescent table lighting is uniform light distribution. Although simple screw-in, capsule or triple twin-tube retrofits can provide adequate illumination, the lighting distribution can be inadequate, especially below the lamp shade. For maximum uniformity of illumination from table lamps, consider installing either 2-D or spiral-shaped retrofit compact fluorescent lamps. These lamps can uniformly brighten the shade, efficiently direct light down to visual tasks and provide uplighting for a pleasing aesthetic effect.

Outdoor Accent Lighting

Line-voltage and low-voltage halogen lighting systems have been used for providing garden and landscape lighting. However, with the increased availability of low-temperature compact fluorescent systems, these applications can be efficiently served with compact fluorescent lamps and low-temperature electronic ballasts (for starting at tempera-

Figure 5-10. Table lamps can be retrofit with either 2-D or spiral-type compact fluorescent lamps, using electronic ballast adapters for instant starting, light weight and improved efficacy. These lamps provide high color rendering (82-86 CRI) and uniform lighting. *Courtesy: Alta Illumination Co.*

tures as low as -20°F to 0°F). To maintain high efficacy in cold outdoor conditions, consider using compact fluorescent lamps with a mercury amalgam for maximum light output and efficacy over a wide range of temperature extremes.

Where longer light projection distances are required (such as for tree or facade uplighting) use point sources such as high-intensity discharge or tungsten halogen. Select the beam spread—from narrow spot to wide flood—that limits the illumination to the highlighted areas.

Chapter 6

Compact Lighting Upgrade Options

T he primary objective for most compact lighting upgrades is to replace standard incandescent lamps with more energy-efficient sources. As new compact lighting products are introduced each year, the incandescent lamp is giving up more of its market share to halogen, compact fluorescent and compact HID. There are few situations where continued use of standard incandescent lamps can be justified.

COMPACT FLUORESCENT LAMP UPGRADES

Compact fluorescent lamps (CFLs) are an energy-efficient, long-lasting substitute for the incandescent lamp. As described in Chapter 5, they are available in a wide variety of configurations beyond the most common twin-tube, quad-tube and triple-twin-tube configurations. CFLs can be purchased as self-ballasted units or as discrete lamps and ballasts. Several retrofit adapters are available for convenient retrofit in existing incandescent sockets. Most CFL products are manufactured with electronic ballasts which provide up to 20 percent higher efficacies as well as instant starting, reduced lamp flicker, quiet operation, smaller size and lighter weight.

Application Guidelines
CFLs may be used in a variety of incandescent applications as shown in Figure 6-1. Also, refer to Chapter 5 for more detail regarding compact fluorescent upgrade applications.

	Recessed Downlights	Surface Units	Pendants	Sconces	Floodlights	Exit & Sign Lighting
T-4 Twin-Tube	O	O	O	●	●	●
T-4 Quad-Tube	● (a)	O	O	●	●	O
Globe Shaped Unit	O	O	●	--	--	--
Integral Lamp-Ballast	O	--	●	--	--	--
Reflector Unit	O	--	O	--	--	--
Electronically-Ballasted Integral Unit	O	O	●	O	O	--
Square	●	●	O	O	--	O

Key
● Superior Lamp Choice
O Suitable Lamp Choice
-- Inappropriate Lamp Choice
Note:
a. with conversion kits complete with reflector

Figure 6-1. Application guidelines for compact fluorescent lamps.
Courtesy: California Energy Commission.

Buyer Beware

Because of their high efficiency and long life, compact fluorescent lamps should be among the first choices for upgrading compact lighting systems. Compared with standard incandescent sources, compact fluorescent sources produce four to five times the initial lumen output per watt of electricity.

However, due to the effects of lamp lumen depreciation, reduced optical control, and thermal sensitivity, *one watt* of compact fluorescent lighting may only replace *three watts* of incandescent lighting to maintain the existing illumination. And operational limitations that affect starting, dimming, power quality and visual comfort should be evaluated before investing in compact fluorescent technology.

Lamp Lumen Depreciation

Because CFLs last significantly longer than standard incandescents, they have more time for their light output to decline as they age. At the end of the compact fluorescent lamp's life, it may have lost 15-20 percent of its initial light output. Before purchasing large quantities of CFLs, ask the manufacturer for independently-measured lamp lumen depreciation data, and base illuminance calculations on the lumens produced when the lamp reaches 40 percent of its rated life. Refer to Chapter 18 for more information about lamp lumen depreciation.

Reduced Optical Control

Because compact fluorescent lamps are not point sources (like incandescents or HID lamps), CFLs are not as effective in projecting light over distance. The light output from a CFL is much more diffuse, and lumens easily stray from the intended target in directional lighting applications. As such, these lamps may not be suitable in high-ceiling downlighting applications (ceilings higher than 15-20 ft.) or where tight control of beam spread is necessary. Note, however, that improvements in CFL reflector design are introduced each year. Perform a trial installation to verify CFL performance in high-ceiling areas.

Dimming

Compact fluorescents may be dimmed, but a new luminaire and/or special ballast is usually required. In general, 2-pin or screw-in compact fluorescent lamps cannot be safely dimmed using conventional incandescent dimming controls. However, dimmable, retrofit (screw-in) compact fluorescents have entered the marketplace.

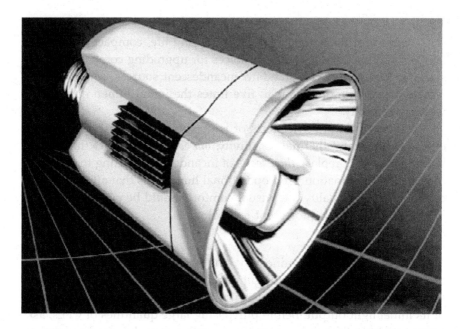

Figure 6-2. With advanced reflector designs, manufacturers are producing compact fluorescent reflector lamps with improved center beam candlepower, suitable for higher mounting heights (over 15 ft.). *Courtesy: ProLight.*

Minimum Starting Temperature

Some CFLs have difficulty starting when the ambient temperature drops below 40°F, while others are designed to start at temperatures well below freezing. Refer to manufacturer specifications for minimum starting temperature.

Sensitivity To Ambient Temperature

The light output of a CFL is significantly reduced when used in a luminaire that traps heat near the lamp (e.g., unvented incandescent downlight) or when exposed to cold temperatures (e.g., outdoors). As a result, CFL upgrades in these common applications can yield disappointing reductions in light output. However, when a mercury amalgam is included in the lamp's chemistry, the light output at temperature extremes is typically within 85 percent of maximum.

Lamp Orientation Effects

In addition, the orientation of the lamp can also significantly affect lumen output. Depending on the lamp design and ambient temperature, the light output in the base-down orientation may be over 15 percent less than in the base-up position. Twin-tube and multiple twin-tube/U-tube designs are the most susceptible to orientation-related light loss. However, the use of amalgam in the lamp can reduce the light loss due to orientation effects. Trial installations are recommended before purchasing large quantities.

Glare Control

The most basic CFL retrofit is to screw a bare CFL into a standard incandescent downlight. Since the fixture's reflector was designed for a concentrated point source (not the longer, more diffuse CFL), much of the light remains trapped in the fixture. Moving the lamp farther down in the fixture will get more light out of the fixture, but will also create glare. The highest performing screw-in retrofit CFL lamps for downlighting applica-

Figure 6-3. This 28W lamp contains four twin tubes and is designed to be used in applications normally served by 100W general-service incandescents. *Courtesy: GE Lighting, Cleveland.*

tions incorporate a reflector specifically designed for the compact fluorescent lamp. To minimize glare, use reflectorized compact fluorescent lamps with a relatively short overall length. Refer to Chapter 5 for additional guidelines to follow when upgrading downlights.

Figure 6-4. These helical-shaped lamps virtually match the size and shape of an A21 incandescent lamp. The increased outside surface area increases light output compared to other compact fluorescent sources of similar dimensions. *Courtesy: GE Lighting, Cleveland.*

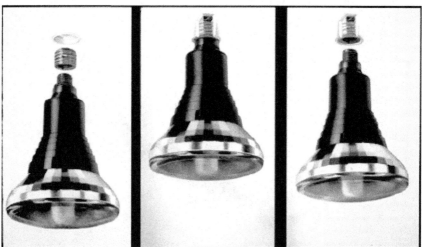

Figure 6-5. Compact fluorescent manufacturers have developed several methods to discourage theft and eliminate "snap-back." For example, this socket will now accept only compact fluorescent lamps with this unique base configuration. *Courtesy: Lumatech Corporation.*

COMPACT HALOGEN LAMP UPGRADES

Compact halogen lamps consist of a small tungsten-halogen cap-sule lamp within a standard lamp shape similar to PAR lamps or general service A-type lamps. These lamps are adapted for use as direct replacements for standard incandescent lamps. Halogen lamps are more efficient, produce a whiter light and last longer than conventional incandescent lamps.

Application Guidelines

As a general rule, compact halogen lamps should be considered for replacing incandescents wherever the more efficient compact fluorescents would not be a better choice. (See the qualifications listed under CFLs above.) Compact halogen lamps can be dimmed, their performance is independent of temperature and orientation, they project light efficiently over long distances, and they present no power quality or compatibility concerns.

The best applications are in accent lighting and retail display lighting, especially where tight control of beam spread is necessary. Other good applications include high-ceiling downlighting and "instant-on" floodlighting.

The use of an optional infrared (IR) coating applied to the halogen capsule can increase the efficacy of this light source by about 35 percent. Both PAR lamps and general service A lamps are now manufactured using this thin film technology.

Halogen directional lamps are available in a wide range of distribution patterns, ranging from very narrow spot lights (as low as 3° beam spread) to wide floodlights (up to 60° beam spread). Vendors of these products can supply easy-to-use guides for determining the required wattage and beam spread needed to deliver the desired brightness in display lighting applications. Chapter 15 provides more detail on retail display lighting.

Buyer Beware

Even though halogen lamps are generally more energy-efficient than standard incandescent lamps, they are only moderately efficacious. Because most compact halogen lamps have an efficacy of around 20 lm/W, they should not be used in applications where compact fluorescent lamps (with efficacies exceeding 40 lm/W) would serve satisfactorily.

Although quartz capsules allow emissions of ultraviolet (UV) light, most compact halogen lamps are equipped with a glass cover or enclosure that blocks nearly all of the UV emissions. Note, however, that some compact halogen task lights, low-voltage halogen lamps and linear quartz lamps may not be equipped with adequate UV protection.

The extended lamp life promised by halogen technology can be compromised if the lamp is dimmed for a prolonged period of time. Under dimming conditions (typically less than 30 percent full output) the halogen cycle stops, and the evaporated tungsten is not redeposited on the filament. When this occurs, the walls of the quartz capsule can become darkened, which increases lamp lumen depreciation. However, these effects can be offset if the lamp is operated for at least 15 minutes on full power.

Bare quartz halogen lamps require special handling, as quartz materials are extremely sensitive to oils and dirt from human skin. Handling of quartz lamps with bare hands can result in bulb wall deterioration and premature lamp failure.

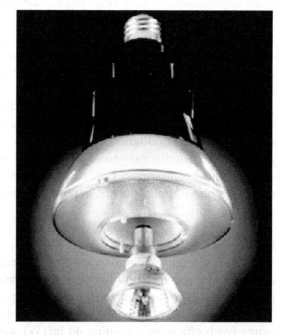

Figure 6-6. Retrofit adapters—with 12V transformers—can be used to convert a line-voltage (120V) socket to one that will accept a low-voltage halogen lamp. These upgrades should be considered where the improved beam control of low-voltage halogen lamps is desired. *Courtesy: C.E.W. Lighting, Inc.*

HARDWIRE COMPACT FLUORESCENT DOWNLIGHT RETROFITS

Instead of simply screwing in a retrofit compact fluorescent lamp in place of an incandescent lamp, the downlight itself can be permanently modified to accept dedicated compact fluorescent lamps.

There are several reasons to consider hardwire downlight modifications instead of simple lamp replacements. The most important reason is to minimize glare from the compact fluorescent source. Downlights are specifically designed to deliver extremely high visual comfort by shielding high-angle illumination. Compact fluorescent downlight modifica-

(mm) dimensions

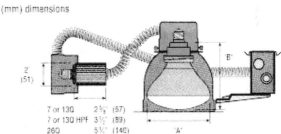

| 2" |
| (51) |

7 or 130	2¼" (57)
7 or 130 HPF	3½" (89)
260	5½" (140)

Figure 6-7. This patented "cap retrofit" can transform many types of incandescent downlights into dedicated compact fluorescent downlights. The existing incandescent downlight cap (that holds the socket) is removed and replaced with the new compact fluorescent cap; this cap comes pre-wired to the compact fluorescent ballast which screws into the original cap for power. The modular CFLs used in this retrofit are enclosed in a reusable R-lamp envelope. *Courtesy: Janmar Lighting.*

tions enable the lamps to be mounted higher or horizontally in the downlight, compared to simple vertical screw-in retrofit lamps.

Another important reason to consider installing a downlight retrofit kit is that the new lamp sockets will only accept dedicated compact fluorescent lamps. So when the lamps burn out, screw-in incandescents cannot be installed by mistake.

Finally, hardwire compact fluorescent downlight conversions yield a higher luminaire efficiency than screw-in retrofit lamps. This efficiency improvement is achieved with the lamp-specific reflector design. Once retrofitted, fixture efficiencies range from 50 to 80 percent.

Application Guidelines

Most hardwire downlight retrofit kits use the existing downlight's mounting frame and junction box, to which new sockets, housing and ballast are attached. In addition, these kits feature metal reflectors specifically designed for the compact fluorescent lamp(s). As the name implies, this retrofit can be permanently installed using a hardwire connection; alternative designs are available that allow the power connection to be made using a screw-in connection in the plenum above the downlight reflector. Contact downlight manufacturers for help in identifying compatible retrofit kits.

Buyer Beware

Due to the diffuse nature of compact fluorescent lighting, compact fluorescent downlight retrofits will typically reduce visual comfort compared to incandescent downlighting. However, hardwire conversions or dedicated CFL downlights provide the best visual comfort when using compact fluorescent lamps.

REPLACE WITH NEW COMPACT FLUORESCENT LUMINAIRES

As the compact fluorescent industry continues to grow, an increasing number of compact fluorescent luminaire designs are being introduced. Compact fluorescent models should be considered for downlights, task lights, table lamps and many other compact lighting applications.

Application Guidelines

Electronic CFL ballasts are available that can operate up to three compact fluorescent lamps at one time. This ballast technology opens the door to a wide range of compact luminaires to suit many applications previously reserved for incandescent luminaires.

When selecting new CFL luminaires, consider the optical design. The luminaire efficiency of a single-lamp luminaire, as compared to a multilamp system of the same total wattage, will inherently be higher because the single lamp system has reduced shadow effects inside the luminaire. For example, a more efficient downlight would consist of a single vertically mounted 26W quad-tube, compared to one that uses two 13W twin tubes.

Another characteristic of new CFL luminaires is their "thermal" performance. With effective venting, new CFL downlights will enable the compact fluorescent lamp to operate closer to its optimal temperature, thereby improving efficacy and light output.

Buyer Beware

When replacing incandescent luminaires with compact fluorescent models, have a lighting professional determine the new layout and calculate the impacts on maintained illuminance, uniformity and visual comfort.

REPLACE WITH COMPACT HID LUMINAIRES

New compact HID luminaires offer the opportunity to improve energy efficiency while maintaining the visual comfort and sparkle of incandescent point sources. With compact HID electronic ballasts available, there are many viable choices for replacing incandescent luminaires with either metal halide or white high-pressure sodium luminaires.

Application Guidelines

The most common applications for compact HID luminaires are in downlighting, floodlighting and display lighting. Because of their point source geometry, these luminaires are suitable for relatively high mounting heights.

Buyer Beware

HID systems have limitations regarding warm-up time, restrike time and dimming. These limitations are described in Chapter 7.

Note that some metal halide sources experience color shift between lamps and over life. Before purchasing these systems, request technical information regarding color performance. Electronic HID ballasts offer improved color stability compared to the electromagnetic models.

Chapter 7

High-Intensity Discharge (HID) Equipment

H igh-intensity discharge (HID) lighting systems offer several distinct benefits, including relatively long life, high efficacy and compact size. Originally developed for outdoor and industrial applications, HID systems are also used in office, retail and other indoor applications. Their color rendering characteristics have been improved and lower wattages are available.

HID LIGHTING SYSTEMS

HID systems are most commonly used for lighting large areas. Due to their generally high lumen output and their point source optics, HID systems can be mounted high above the task area to illuminate large areas—using a minimum number of luminaires.

HID lamps are similar to fluorescent lamps in that an arc is generated between two electrodes. However, the arc in an HID source is shorter, and it generates much more light, heat and pressure within the arc tube. Included in the family of HID light sources are mercury vapor, metal halide and high-pressure sodium. Because low-pressure sodium lamps are considered to be low-pressure discharge sources (like fluorescent), they are not technically considered to be HID lamps. Nevertheless, this chapter also addresses low-pressure sodium lamps because they are related to the high-pressure sodium lamp.

MERCURY VAPOR LAMPS

Mercury vapor lamps, which produce a blue-green light, consist of a mercury-vapor arc tube with tungsten electrodes at both ends.

These lamps have the lowest efficacies of the HID family, rapid lumen depreciation and a low color-rendering index (as low as 15). Because of these characteristics, other HID sources have replaced mercury vapor lamps in most applications. The arc is contained in an inner bulb called the arc tube, which is filled with high purity mercury and argon gas. This arc tube is enclosed within the outer bulb, which is filled with nitrogen. Color-improved mercury lamps use a phosphor coating on the inner wall of the outer bulb to boost the CRI rating to 50.

METAL HALIDE LAMPS

The construction of metal halide lamps is similar to that of mercury vapor lamps. However, the arc tube contains metal halide additives in addition to the mercury and argon. These additives enable the lamp to produce more visible light per watt with improved color rendition. Metal halide lamps can be identified by their white light with color temperatures of 3000-5000K and typical CRI ratings of 65-70. New metal halide lamps with ceramic arc tubes can replace high-wattage incandescent lamps in many applications. Producing warm color tones with high color rendering and high efficiency, these sources are becoming the standard for use in high-quality, energy-efficient compact luminaires.

With a wide range of wattages—under 35 watts to over 1,500 watts, metal halide lamps can be used in many indoor and outdoor applications. Because of their good color rendition and high lumen output, these lamps are useful for many applications, including sports arenas and service stations. Typical indoor uses include large auditoriums and high-ceiling retail applications. Metal halide systems should be specified in large-area or directional lighting applications where color rendering *and* energy efficiency are important.

Defining Characteristics of Metal Halide Lamps
Below is a list of defining characteristics of metal halide lamps:

High Efficiency
The maintained efficacy of metal halide lamps ranges from 40 to 90 lm/W—typically about double that of mercury vapor.

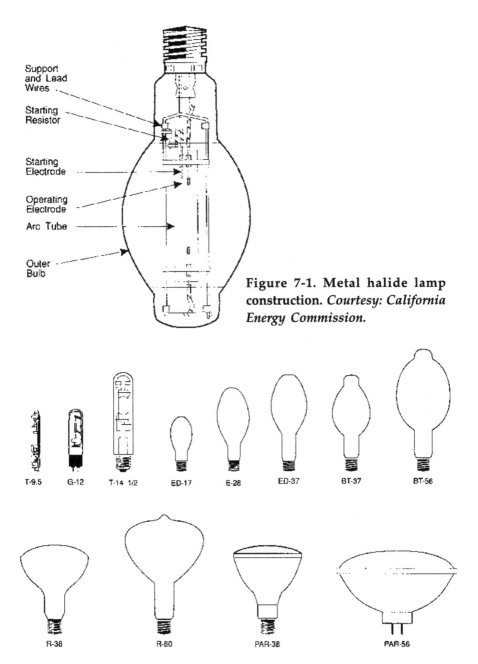

Figure 7-1. Metal halide lamp construction. *Courtesy: California Energy Commission.*

Figure 7-2. Common metal halide lamp shapes. Courtesy: California Energy Commission.

Color Rendering

Metal halide lamps provide the highest degree of color rendering of all standard HID lamps. The standard clear metal halide lamp has a CRI rating of 65. Alternatively, the phosphor coating on metal halide lamps improves the CRI rating to 70 and provides more uniform brightness, a warmer appearance and softer shadows. The low-wattage metal halide lamps with ceramic arc tubes produce a CRI exceeding 80.

Color Shift and Variation

The various chemicals inside the lamp arc tube produce different colors. As the metal halide lamp ages, chemical changes occur in the lamp causing shifts in lamp color temperature—by as much as 600K. However, metal halide lamps with ceramic arc tubes deliver improved color stability, maintaining the color temperature within 150K. In addition, the use of electronic metal halide ballasts can reduce color shift.

Operating Position

Certain varieties of metal halide lamps are designed to operate in a specific burning position, such as horizontal, vertical with base up and vertical with base down. These position-specific lamps provide maximum efficacy and lamp life. Lamp manufacturers usually designate the correct burning positions of their position-sensitive lamps. Operating metal halide lamps in burning positions other than those recommended by the manufacturer will adversely affect lamp life and lumen output.

Safety

It is possible for the inner arc tube to rupture upon lamp failure. The lamp operation and maintenance practices below are recommended to minimize the possibility of arc tube rupture:

•   Relamp metal halide fixtures at or before the end of the rated life. Removing lamps before they are expected to fail will minimize the possibility of arc tube rupture.

•   In applications where the metal halide lamps are in use 24 hours per day, turn the lamps off at least once per week for at least 15 minutes. The power surge from restarting will cause the weakest lamps to immediately fail, presumably under controlled circumstances. Alternatively, use special metal halide lamps that are designed for use with "regulated lag" ballasts, which can be operated for three years—without interruption—before being group relamped.

- The outer envelope shields most of the ultraviolet radiation emitted by the arc tube. However, if the outer envelope breaks, shortwave ultraviolet radiation can cause skin burn and eye inflammation. Certain lamps are designed to automatically extinguish when the outer envelope is broken or punctured. Another type of safety lamp features a shrouded arc tube that protects the lamp and occupants in the event of an arc tube rupture. These safety lamps are sometimes listed as "open fixture rated"; other standard metal halide lamps must be used in enclosed HID luminaires.

Specialty Metal Halide Lamps
Below is a list of the many types of metal halide lamps available for specific uses:

Universal-position Metal Halide Lamps
Although universal-position metal halide lamps can be operated in any position, they generally perform best when the arc tube is in a vertical position, yielding longer life and higher lumen output than when the arc tube is off vertical by more than 15 degrees. To obtain optimum metal halide lamp performance, however, use position-specific lamps whenever the operating position is known.

Vertical-position Metal Halide Lamps
The vertical burning metal halide lamp is specifically optimized for either base-up or base-down operation. The most common application for these lamps is high-ceiling downlighting. The principal advantage of vertical burning lamps is efficacy—these lamps generate about 10 percent more lumens than vertically oriented universal lamps, without using more energy.

Horizontal-position Metal Halide Lamps
Compared to a universal metal halide lamp operated in the horizontal position, these position-optimized lamps produce up to 25 percent greater light output and up to 33 percent longer life. These lamps often have bowed arc tubes, and they use a position-fixing pin in the base

called a position-orienting mogul (POM) base. This base and matching socket assure correct rotational position of the lamp's arc tube for maximum performance. Also known as "high output" or "super" metal halide lamps, these lamps are commonly used for billboard lighting, gas station canopies and indirect lighting.

Horizontal-position metal halide lamps are also available in the double-ended configuration. The principal advantage of this lamp type is its compact size which contributes to improved optical performance in retail display lighting applications. High wattage versions of this lamp are used in sports lighting applications.

Instant and Quick Restrike Metal Halide Lamps

Of all the HID lamps, metal halide lamps take the longest to restrike after a momentary power interruption—up to 20 minutes. During this time, the arc tube must cool down before it can restrike and begin the 2-5 minute warm-up process again. For applications where such a possibility is intolerable (such as lighting of televised professional sports), an immediate restrike lamp is needed. However, because instant-restrike metal halide lamps are relatively expensive and complex, *quick* restrike lamps have been developed. The quick restrike lamps feature a high voltage pulse ignitor (with a compatible socket) which typically enables restrikes within one minute of restored power.

Directional Metal Halide Lamps

The compact size and high CRI characteristics of the metal halide source led to the development of a variety of directional metal halide lamps. Similar to incandescent and halogen directional lamps, metal halide lamps are now available in a wide range of PAR and R lamps.

Metal halide PAR lamps are produced in packages of up to 400 watts and in a wide range of beam spreads—from spot lighting to wide flood lighting. PAR metal halide lamps are also advantageous because they are rated for use in open fixtures; because these PAR lamps do not require a protective luminaire cover, overall luminaire efficiency can be increased.

The R40 metal halide lamps also have optical control built into the lamp. These lamps can be specified as spot or flood lamps with wattages ranging from 70 to 175. These reflector lamps (and the PAR56 and PAR64 lamps) require a glass covering for protection in the event of an arc tube rupture.

One advantage of these directional lamps is that the reflective surface is sealed from the effects of dirt and corrosion, reducing light output depreciation. Because these lamps require a metal halide ballast, retrofits usually involve a complete system replacement.

HIGH-PRESSURE SODIUM LAMPS

The high-pressure sodium (HPS) lamp is widely used for street lighting and industrial applications. With a CRI rating of 22, the standard high-pressure sodium lamp serves well in applications where color rendering is not critical.

The efficacy of the HPS system is very high—up to 115 lm/W, maintained. As shown in Table 7-1, the 400W HPS system produces nearly 2-1/2 times the light output compared with the same-wattage mercury vapor system.

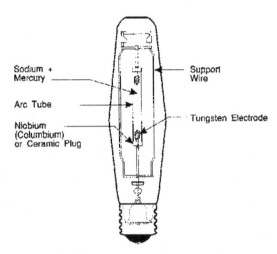

Figure 7-3. High-pressure sodium lamp construction. *Courtesy: California Energy Commission.*

Table 7-1. Lumen output comparison of 400W HID lamps.

HID System	Maintained Lumens
400W Mercury Vapor	18,900
400W Metal Halide	32,000
400W High-pressure Sodium	45,000

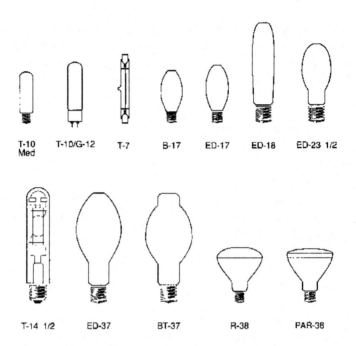

T-10   T-10/G-12   T-7   B-17   ED-17   ED-18   ED-23 1/2
Med

T-14 1/2   ED-37   BT-37   R-38   PAR-38

Figure 7-4. Common high-pressure sodium lamp shapes. *Courtesy: California Energy Commission.*

HPS lamps differ from standard mercury and metal-halide lamps in that they do not contain starting electrodes; the ballast circuit includes a high-voltage electronic starter. The arc tube is made of a ceramic material which can withstand temperatures up to 2,372°F. It is filled with xenon to help start the arc, as well as a sodium-mercury gas mixture. Sodium, the major element used, produces the 2100K "golden" color temperature that is characteristic of standard HPS lamps.

Universal Position High-pressure Sodium Lamps

HPS lamps operate independently of orientation. In addition, HPS lamps do not require an enclosure, except to prevent moisture from accumulating on the lamp. Therefore, there are far fewer types of HPS lamps than metal halide.

Improved Color Rendering HPS Lamps

Improved color properties are available in the "deluxe" and "white" HPS versions which provide higher color temperature and improved color rendition. However, these color improvements are achieved by sacrificing efficacy.

Deluxe HPS Lamps

While remaining compatible with standard HPS ballasts, this replacement lamp will dramatically improve color rendering—from the standard HPS lamp's 22 to a respectable 60-65. However, this retrofit will result in a loss of over 20 percent in maintained light output, with a corresponding drop in efficacy. The primary difference between deluxe HPS and metal halide is the color temperature: 2200K for deluxe HPS vs. 3700K for standard metal halide.

White HPS Lamps

White HPS lamps closely resemble the appearance of incandescent lamps, both in color rendering and in color temperature. Although the efficacy is relatively low (about the same as mercury vapor), it exceeds the efficacy of halogen lamps.

Directional Lamps

HPS lamps can be purchased in the R38 and PAR38 envelopes for providing wide flood lighting (65° beam spread). Directional HPS lamps are available for both standard and deluxe HPS sources for general floodlighting and track lighting applications, respectively. And PAR36 white HPS lamps may be used for spot and flood applications.

Instant-restrike HPS Lamps

Some HPS lamps are produced with two arc tubes that provide instant (or "standby") restrike cycles while offering extended lamp life. Although there will still be a warm-up time following a power interruption, the lamp will not have to cool down before the second arc can be struck. In normal operation, these lamps alternate operation between the arc tubes. Note that after one arc tube fails, the HPS lamp will continue to operate, but without the standby restrike capability. Instant-restrike HPS lamps are available in wattages ranging from 70 to 1,000 watts.

LOW-PRESSURE SODIUM LAMPS

Low-pressure sodium (LPS) lamps are the most efficient light sources, but they produce the poorest quality light of all the lamp types. Being a *monochromatic* light source, all colors appear black, white, or shades of gray under an LPS source—the CRI rating of the LPS lamp is zero. LPS lamps are available in wattages ranging from 18 to 180 and are easily identified by their "pumpkin-orange" color.

LPS lamp use has been generally limited to outdoor applications such as security or street lighting, and in indoor, low-wattage applications where color quality is not important (e.g., emergency stairwells). However, because the color rendition is so poor, many municipalities do not allow their use in roadway lighting.

LPS lamps are particularly useful for outdoor lighting in the vicinity of observatories. Because the LPS lamp emits light within a narrow band of wavelengths, the astronomers can use filters on their telescopes to eliminate any reflected LPS light and thereby maintain a very clear view of the heavens.

Because of the relatively long shape of the LPS lamp, it is not as optically efficient as a point source would be in directing and controlling a light beam. Therefore, its use is generally limited to mounting heights under 25 ft.

Another special characteristic of the LPS lamp is the volatile nature of sodium when it comes in contact with water. LPS lamps should only be disposed of by qualified contractors.

HID BALLASTS

Like fluorescent lamps, HID lamps require a ballast to provide the necessary starting voltage, limit the operating current and regulate the voltage supplied to the lamp. With increasing use of HID lighting systems, a variety of new HID ballast types have emerged. The demand for

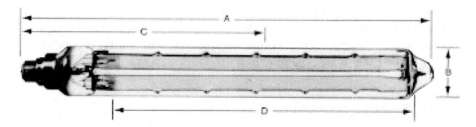

Figure 7-5. Low-pressure sodium (LPS) lamps produce light in a U-shaped discharge tube. *Courtesy: Philips Lighting Company.*

these new ballasts has been driven by needs for longer lamp life, higher lumen output, quicker starting, increased efficacy and improved color stability.

Although the newer HID ballasts are more energy efficient, the gains in efficacy are rarely enough to cost-justify a ballast retrofit. Generally, the cost-effective applications of improved HID ballasts include system conversions (e.g., converting a mercury system to metal halide), new construction, renovation projects and replacement of failed HID ballasts.

Magnetic HID Ballasts

Most HID ballasts use magnetic coils to regulate current. Operating at 60 Hz, these large, heavy ballasts produce a noticeable hum unless they are encased and potted. The total harmonic distortion for the typical magnetic HID ballast is less than 30 percent. There are several types of magnetic HID ballasts, as described below.

Constant Wattage Auto Transformer (CWA)

The CWA ballast is the most common type of HID ballast for wattages of 175 and higher. These ballasts typically provide high power fac-

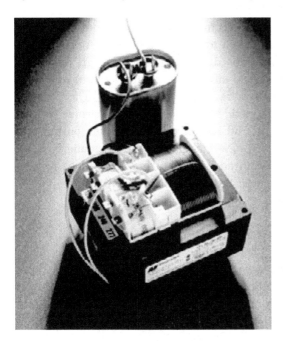

Figure 7-6. Most HID ballasts have a capacitor that is separate from the rest of the ballast components. *Courtesy: Magnetek Lighting Products Group.*

tor (>0.90) and can tolerate a voltage drop of 30 percent without the lamp extinguishing.

Reactor Ballast (R)

The reactor ballast is the simplest type of HID ballast. Although these ballasts have low internal losses, they typically have a low power factor (0.50) and can cause lamps to flicker or shut off if the voltage varies.

High Reactance Auto Transformer (HX)

Most commonly used with low-wattage (<100W) HID lamps, HX ballasts are similar to reactor ballasts, but they can boost line voltage as needed to start the lamps. Electronic HID ballasts are becoming a popular alternative to the HX ballast.

Electronic HID Ballasts

Now being introduced at an ever-increasing rate, electronic ballasts for low-wattage metal halide and HPS lamps provide high power factor and lower energy losses than magnetic ballasts. Future electronic HID ballast developments will yield ballasts for use with higher wattage HID lamps. In most electronic HID ballast designs, the total harmonic distortion is less than 10 percent.

Although HID systems with electronic ballasts are more efficacious, the boost in efficacy is not sufficient to cost-justify system retrofits. Unlike fluorescent lamps, HID lamps do not respond to high frequency power with significantly increased efficacy. The improvement in energy efficiency is mostly the result of reduced ballast losses. The primary benefits of replacing failed magnetic HID ballasts with electronic ballasts are smaller size, lighter weight and better control of lumen output and color temperature over time. In new installations, their smaller size allows them to be directly mounted into fixtures—instead of remote mounting—producing savings on the initial installation cost.

PERFORMANCE CHECKLIST FOR HID SYSTEMS

There are wide differences in performance between each member of the HID lamp family. In addition, the use of new types of HID ballasts can improve the performance of HID lighting systems. Table 7-2 outlines the performance characteristics of HID lighting systems.

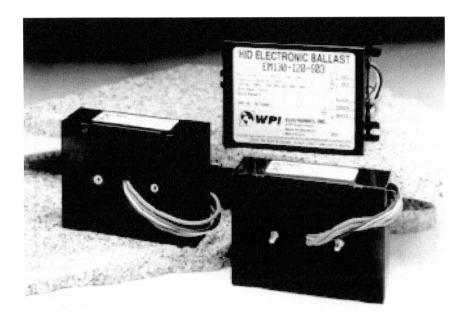

Figure 7-7. Electronic HID ballasts offer improved efficacy, smaller size, lighter weight and improved lamp color consistency. *Courtesy: WPI Electronics, Inc.*

Energy Efficiency Versus Color Rendering

The most efficacious HID sources are those with relatively low color rendering performance. Therefore, select the most efficient light source that meets the minimum requirements for color rendering. Note that metal halide and deluxe high-pressure sodium systems provide similar levels of efficacy and color rendering.

Warm-up Time

It is not possible to instantly ignite a cold HID lamp and produce full brightness. As power is applied, the temperature and pressure inside the arc tube builds gradually, causing metallic vapors to enter the arc and release light energy. The duration of the warm-up period varies depending on the lamp type, ranging from 2 to 10 minutes. During this period, the lamp will exhibit different colors as the various metals vaporize. Metal halide and HPS lamps typically require at least 2-3 minutes to warm up to full brightness.

Table 7-2. Typical HID system performance. *Source: Manufacturer literature.*

Lamp Type	Wattage	Color Temperature	CRI	Relative Efficacy[1]	Lamp Life (hrs x 1000)	Warm-up Time (min.)	Restrike Time (min.)
MERCURY VAPOR							
Clear	100-1000	5700K	15	100% (base)	24+	5-7	3-6
Phosphor-Coated	50-1000	3300-3900K	50	105%	24+	5-7	3-6
METAL HALIDE							
Clear	35-1500	3200-5200K	65	200%	5-20[2]	2-4	5-15
Phosphor-Coated	32-400	2700-4000K	70	195%	5-20[2]	2-4	5-15
Ceramic Arc Tube	35-150	3000K	83	220%	7.5-10	3-5	4-8
HIGH-PRESSURE SODIUM							
Standard	35-1000	2100K	22	250%	16-24	3-4	1-2
Deluxe	70-400	2200K	65	205%	18-20	3-4	1-2
White HPS	35-100	2600-2800K	85	100%	14-18	3-4	1-2

[1]Approximate relative lumens per watt compared to mercury vapor systems of similar wattage.
[2]Shorter lamp life values apply to lower-wattage lamps.

Lamp Restrike Time

The HPS lamp has the most rapid restrike time of the HID lamps: about one minute. Although standard metal halide lamps operating on a CWA ballast can take up to 20 minutes to restrike, the use of alternative ballasts (such as reactor, high reactance and electronic ballasts) will provide restrike times of less than four minutes. In applications where a brief power outage could cause hazardous conditions or a manufacturing shutdown, and where there are no instant-on backup emergency lighting systems in place, it is a good idea to specify that some of the luminaires provide either instant-restrike capability or halogen backup lamps.

Temperature Sensitivity

Although metal halide lamps start at temperatures as low as –20°F, their life may be reduced if they are frequently started below 10°F. HPS lamps are fairly insensitive to temperature and will start at temperatures as low as –40°F. Contact the lamp supplier for specific guidelines for use in cold weather.

Color Shift and Variation

As noted earlier, metal halide lamps are susceptible to color variations between similar lamp types, and the color temperature can shift over the life of the lamp. In addition, the improved-CRI HPS lamps can gradually lose their improved CRI performance over their lamp life. The use of electronic HID ballasts helps to stabilize color shift and variation in HID lamps.

Dimming

HID lamp output can be adjusted with the use of dimming electronic HID ballasts, using capacitive switching (step-dimming) ballasts, or with a panel-level dimming system. Panel-level dimming systems reduce power to the circuit by reducing voltage or current, or by modifying ("chopping") the voltage waveform.

Operating HID lamps at reduced output will produce a shift in color and will reduce lamp efficacy. For example, a metal halide lamp can be dimmed to about 40 percent power, but at this level, it generates only about 25 percent of its rated lumens. In addition, as the lamp dims, the color temperature increases (becomes bluer) and its color rendering performance decreases. Coated metal halide lamps provide a better color appearance when dimmed than clear lamps. Note that continuous dimming at low power levels will reduce lamp life and increase lamp lumen depreciation.

Lamp Life

The rated life of metal halide lamps is shorter than other HID sources. Low-wattage lamps last less than 7,500 hours while high-wattage lamps last an average of 15,000-20,000 hours.

Lamp Failure Mode

When HPS lamps reach end of life, they exhibit a unique mode of failure. The lamp will start, build in light output and go out. This cycle will be repeated at 1- to 2-minute intervals until the lamp is replaced or the ignitor has failed. Therefore, it is important to replace failed HPS lamps as soon as possible to prevent premature ignitor failure. Ignitor replacements can exceed $150, including labor. HID lamp manufacturers now offer HPS lamps that include the ignitor system *inside the lamp,* rather than inside the ballast compartment. Each time the lamp is replaced at end of life, the ignitor is also replaced, eliminating potential ignitor problems caused by cycling. As an alternative, special HPS lamps are designed to change to a blue color at end of life, while eliminating the end-of-life cycling characteristic.

Strobe Effects

All HID lamps are turned on and off 120 times per second in 60-Hz alternating current power circuits. The use of HPS lamps near rotating machinery may produce a stroboscopic effect, making the machinery appear to be motionless—a potentially hazardous situation. This can occur when the moving object rotates at any speed that is a multiple of 60 (i.e., 2,400 rpm). Strobe effects of this type can be minimized by the proper phasing of the luminaire power supply circuits, so that none of the machinery is lighted solely by luminaires on the same phase circuit. Alternatively, metal halide lamps can be used, because they do not create strobe problems.

TYPES OF HID LUMINAIRES

The general categories of HID luminaires are identified by their application, such as high-bay industrial, floodlighting, commercial outdoor and roadway. Common to all these luminaires is the use of optical and reflector assemblies for delivering a specific lighting distribution pattern. One of the key goals in HID luminaire selection is to choose luminaires with distribution patterns that are relatively rectangular in order to deliver uniform illumination using a minimum number of luminaires.

High-bay Indoor HID Luminaires

Most indoor HID lighting is divided into either high-bay luminaires (generally for mounting heights over 15-20 ft.) or low-bay luminaires (generally for mounting heights under 15-20 ft.). The selection of high-bay luminaires is driven by the type of lamp used and the dirt conditions expected in the space. HID luminaires are chosen to maximize task visibility (for safety and productivity), while minimizing luminaire dirt depreciation (for maintaining luminaire efficiency).

Clear Reflectors

In *relatively clean environments*, high-bay luminaires with clear reflectors should be used. These luminaires distribute most of the light down to the floor, but they also allow a portion of the light to be directed up to the ceiling where it is reflected off of a light-colored surface. This indirect lighting component diffuses the light and enhances the illumination on vertical surfaces. Because many of the tasks in industrial applications are performed in three dimensions (as opposed to most office applications that are confined to two-dimensional work surfaces), illumi-

Clear Prismatic (clean environment)   Open Ventilated (moderately dirty environment)   Enclosed Gasketed (very dirty environment)

Figure 7-8. The selection of high-bay HID luminaires is driven by the type of dirt conditions in the space. In addition, some metal halide lamps must be used in enclosed fixtures; others are rated for use in open fixtures. *Courtesy: EPA Green Lights.*

nating vertical surfaces will help improve task visibility. Because these luminaires utilize the ceiling for diffusing light, it is important to keep the ceiling clean and painted with a highly reflective white paint in order to maintain high system efficiency.

Open Ventilated Fixtures

In *moderately dirty environments*, clear reflectors would not be a wise choice because dirt accumulation on room surfaces would absorb much of the uplighting. Instead, the high-bay HID system in these conditions should consist of opaque, ventilated fixtures which allow heat-generated air currents to flow upwards through the luminaires. This air flow reduces the effect of luminaire dirt depreciation because the airborne dirt particles are carried through the luminaire instead of adhering to the lamp and optical surfaces.

Enclosed Gasketed

In *very dirty environments*, enclosed gasketed luminaires inhibit the entrance of airborne dirt particles, keeping the lamp and reflector relatively clean. However, the luminaire's lens must be cleaned frequently to maintain high luminaire efficiency. The gasketing material must be continuous and permanently attached to one surface so that it will not come loose or fall off during luminaire maintenance. Silicone gaskets generally are more expensive, but because silicone does not permanently deform with use, these gaskets usually maintain their effectiveness.

Enclosed Metal Halide Luminaires

Where standard metal halide lamp operation requires an enclosed fixture for safety, the luminaire may or may not be gasketed. In non-gasketed luminaires, dirt can build up on the inside surface of the lens. Because the lens absorbs light, these luminaires are less efficient than open luminaires. Therefore, unless the fixture is operating in a very dirty environment (where an enclosed gasketed luminaire should be used), we can improve efficiency by specifying an open luminaire and a metal halide lamp that is rated for open fixture operation.

Low-bay Indoor HID Luminaires

The challenge of providing low-bay HID lighting is to achieve a wide angle of distribution (for uniformity) while shielding occupants from direct glare. Because of their relatively low mounting height, these luminaires are frequently located within the occupants' normal field of view. Therefore, it can be difficult to provide wide-angle distribution without causing some glare. The most common low-bay HID luminaires feature an acrylic "low-glare" diffuser. The term "low-glare" simply means that the intensely bright HID arc tube is shielded by the diffuser. The diffuser itself can be relatively bright, resulting in low visual comfort. Some low-bay luminaires have a sharp cut-off angle for high visual comfort, but the spacing criteria are reduced, requiring more luminaires for uniform lighting.

Outdoor HID Luminaires

There are many outdoor lighting applications that utilize the unique operating advantages of HID sources.

HID Floodlighting

Floodlighting luminaires are specified by their beam spread in degrees, for both horizontal and vertical spreads. HID floodlights may be used for facade lighting, outdoor signs, outdoor sports lighting and general security lighting. These luminaires may be mounted on walls or poles.

Figure 7-9. Low-bay HID luminaires must provide wide-angle light distribution while diffusing or shielding the arc tube's intense brightness. *Courtesy: Thomas Lighting, Inc.*

Roadway Lighting

To illuminate roadways, the luminaires are specifically designed for throwing a long, narrow beam of light on the roadway. Street lights are classified into different types based on their lighting distribution pattern and their lateral "throw" distance. The most common type of street light is the refractor (or cobrahead) luminaire.

Parking Lot Lighting

Parking lot luminaires are commonly pole-mounted in various configurations. Options for parking lot luminaire mounting include either post-top or a combination of one or more luminaires positioned on an arm off to the side of the pole.

Refer to Chapter 17 for more application guidelines regarding outdoor lighting.

Chapter 8

High-Intensity Discharge (HID) Upgrade Options

F or years, the choices for HID lighting upgrades were limited to metal halide and high-pressure sodium system conversions. Today, however, there are several opportunities for improving energy efficiency and/or lighting quality with new HID components and systems. Investments in reduced-wattage HID lamps, retrofit reflectors, HPS lamps for mercury ballasts and bi-level HID luminaire switching can yield substantial profits in applications where high intensity discharge lighting is used.

ENERGY-SAVER HID LAMPS

Reduced-wattage metal halide and HPS lamps are available from most HID lamp manufacturers that reduce energy consumption by up to 18 percent with corresponding reductions in light output. "Energy-saver" versions of metal halide and HPS lamps are available in 225W and 360W packages for directly replacing 250W and 400W lamps, respectively. In addition, 150W metal halide energy-saver lamps are available for replacing the 175W lamp. These retrofit lamps are designed to be compatible with existing luminaires and ballasts.

Application Guidelines
Energy-saver HID lamps are economical upgrades for spaces that are overlighted. And by replacing the lamps at 60-70 percent of their rated life, maintained light levels can actually be *increased* through reduced lamp lumen depreciation, assuming that the current system is relamped upon individual lamp failures. See Chapter 20 for maintenance procedures that can improve system efficacy and lumen output.

Buyer Beware

Because the system efficacy (lumens per watt) is virtually unaffected by this retrofit, the use of energy-saver HID lamps will reduce maintained light levels unless system relamping is performed more frequently to offset the effects of lamp lumen depreciation.

This upgrade to lower wattage HID lamps is subject to snapback. It is possible for the energy savings to cease if full-wattage lamps are used as replacements for energy-saver HID lamps that have burned out.

RETROFIT HID LAMPS

Specially designed HPS and metal halide lamps can be used in place of specific wattages of mercury vapor lamps, without requiring a ballast change.

Application Guidelines

As shown in Table 8-1, these lamps provide an inexpensive alternative for significantly improving light output while saving up to 14 percent in energy consumption in existing mercury vapor luminaires. Note that several manufacturers produce specially designed metal halide lamps that will operate on existing HPS ballasts for improving color rendering (to 65-70 from 22), but causing maintained light output reductions of 33-50 percent.

Buyer Beware

When considering an upgrade with retrofit HID lamps, verify that all components are compatible. For example, confirm that the socket rating is compatible with the new lamp type. Some lamp manufacturers state that their lamps are compatible with specific types of mercury vapor ballasts (such as high reactance autotransformer ballasts) while other manufacturers claim that their lamps may be used with any mercury ballast. Contact the manufacturer for specific application instructions.

Conduct a trial installation to determine if resultant light levels and distribution will be acceptable. If the resultant increase in light output is too great, consider installing the retrofit HID lamps in fewer luminaires. However, check the luminaire's spacing criteria and deter-

Table 8-1. Retrofit HID lamps (no ballast change). *Source: Manufacturer literature.*

Base Lamp Type	Upgrade Lamp Type	Wattage Savings	Lumen Increase*
175W Mercury Vapor	150W HPS	14%	82%
250W Mercury Vapor	215W HPS	14%	97%
400W Mercury Vapor	360W HPS	10%	131%
400W Mercury Vapor	325W Metal Halide	19%	4%
250W HPS	250W Metal Halide	0%	-50%
400W HPS	400W Metal Halide	0%	-33%

*Based on average maintained lumens, accounting for lamp lumen depreciation over lamp life.

Note: All lamps assumed to be clear (uncoated).

mine if the lighting uniformity will be acceptable in cases where every other luminaire contains a lamp. (Refer to Chapter 1 for a complete discussion of spacing criteria and lighting uniformity.)

For greater energy savings and wattage selection, consider replacing the mercury vapor luminaire with a new HPS or metal halide luminaire. New HID luminaires offer opportunities for specifying electronic ballasts, advanced controls and optimized light distribution.

INSTANT-RESTRIKE HPS LAMPS

Instant-restrike HPS lamps contain *two* arc tubes—only one is in use at a time. The term "instant-restrike" is used because if these lamps are switched off, they can immediately be turned back on again, instead of having to wait up to a full minute for the arc tube to cool

Figure 8-1. "Instant-restrike" high-pressure sodium lamps use two arc tubes. The "standby" tube is ready for restriking in the event of a short power interruption. *Courtesy: GE Lighting, Cleveland.*

down before restriking. During normal operation, the second (standby) arc tube remains cool, ready for immediate striking if needed.

Application Guidelines

Instant-restrike HPS lamps can be used in various occupancy sensor applications. For example, instant-restrike HPS lamps can be used as an alternative to bi-level HID systems for illuminating intermittently occupied warehouse aisles. This inexpensive alternative simply involves a lamp change, rather than a system retrofit. As shown in Table 8-2, the lamps will immediately deliver about 11 percent of full light output upon activation and will rapidly rise to full output within three minutes. Instant-restrike HPS lamps are available in wattages ranging from 70-1,000 watts.

Buyer Beware

Although the lumen output rises rapidly upon activation, a trial installation should be performed to verify that sufficient lighting will

Table 8-2. Instant-restrike HPS lamp performance. *Source: Osram Sylvania.*

Time From Power Interruption	Percent Full Light Output
0	11%
1 minute	60%
2 minutes	90%
3 minutes	100%

be delivered to the area when occupants enter the space to perform work.

Note that after one arc tube fails, the HPS lamp will continue to operate, but without the standby restrike capability.

When relamping instant-restrike lamps that are controlled by occupancy sensors, be certain that the replacement lamps are also instant-restrike. Using standard HPS lamps on a circuit controlled by occupancy sensors for on/off control can present a potentially unacceptable time delay between occupancy and restored light levels.

HIGH PERFORMANCE METAL HALIDE SYSTEMS

For maximum metal halide system efficacy, consider a new high-performance system that uses a special metal halide lamp (without an ignitor) and a dedicated pulse-start magnetic ballast with a built-in ignitor. These ballasts are available in both reactor and CWA versions.

The main advantage of this system is the energy reduction from reduced ballast losses, compared to standard ballasts. The bottom-line results of using the 350W system in place of the standard 400W system are a 20 percent savings in energy use, and a 14 percent improvement in efficacy. Other benefits of this system include extending lamp life by up to 50 percent and reducing the hot restrike time to only 2-4 minutes.

Figure 8-2. Pulse-start ballasts for metal halide lamps are designed to operate reduced-wattage lamps while saving 50 percent in ballast losses. Their compact size makes them suitable for retrofit applications. *Courtesy: Advance Transformer Co.*

Application Guidelines

To convert to this system, both the lamp and the ballast must be replaced, not unlike a T8 lamp/ballast conversion. The compact ballasts can be retrofit in place of the existing metal halide ballast. If operating hours are long and electricity prices are relatively high, this conversion can be a cost-effective retrofit. Also, consider this system for maximizing efficacy in new metal halide installations.

Buyer Beware

Common to all reactor ballasts is a low tolerance for voltage dips. Because pulse-start reactor ballasts do not provide any voltage transformation functions, the supply voltage must meet the lamp's voltage requirement. Where improved voltage regulation is needed, specify CWA pulse-start ballasts. When retrofitting with pulse-start ballasts, make sure the lamp sockets are rated to handle the higher starting current.

RETROFIT HID REFLECTORS

Conventional reflectors in open HID luminaires can be retrofit or replaced with specular or clear reflectors in order to enhance luminaire efficiency and control glare.

In relatively clean environments, retrofit HID reflectors can increase illuminance on task surfaces without increasing energy consumption. In overlighted spaces, the efficiency improvement may allow some of the luminaires to be removed, de-energized or relamped with energy-saver (reduced-output) HID lamps. In addition, proper applications of retrofit reflectors can reduce glare and/or improve the lighting distribution characteristics.

Application Guidelines

There are several applications where retrofit HID reflectors may be a wise investment. In situations where the existing reflector surfaces

Figure 8-3. Retrofit specular HID reflectors can improve task illumination without increasing power consumption. *Courtesy: C.E.W. Lighting, Inc.*

are deteriorated and cleaning cannot restore the efficiency of the luminaire, retrofit HID reflectors can be a cost-effective approach for improving luminaire efficiency. In applications where little or no uplighting is desired, specular downlighting reflectors can be installed to deliver a higher percentage of lumens to the horizontal workplane. Retrofit HID reflectors can also be used to modify the lighting distribution characteristics of the base luminaire. For example, retrofit reflectors can modify the existing luminaire's spacing criterion (distribution width), and they can alter the degree of uplighting provided through clear refractors.

Buyer Beware

The installation of retrofit reflectors will alter the lighting distribution from existing HID luminaires. When evaluating a trial installation, check for uniformity of illuminance, visual comfort (glare), illuminance on vertical surfaces, color shift and aesthetic effects.

Figure 8-4. Existing HID luminaires can be retrofitted with clear reflectors which direct most of the light down to the work surfaces. Some of the light is refracted toward the ceiling and walls, providing diffuse reflected light for improving illuminance on vertical surfaces. This product allows flexible lamp positioning which offers a range of spacing criteria and uplighting percentages. *Courtesy: Lexalite International.*

These retrofit products are primarily intended for use in open HID luminaires located in relatively clean environments where the rate of luminaire dirt depreciation will not minimize the efficiency gains achieved by the reflector.

CONVERSION TO NEW HID SYSTEM

Because metal halide and HPS systems are excellent point sources that produce high efficacy, they are good candidates for replacing existing high-bay or outdoor lighting systems that use incandescent, mercury vapor, or (in some cases) fluorescent lamps. These retrofits normally include a complete luminaire replacement, including the lamp, ballast and optical assembly. Refer to the HID equipment overview presented in Chapter 7 for a complete discussion of these lamps and their characteristics.

Application Guidelines

The most cost-effective upgrades involve replacing less-efficient sources such as incandescent, HO/VHO fluorescent, or mercury vapor with metal halide or HPS systems. Upgrades may involve a one-for-one luminaire replacement or a new layout of luminaires to take advantage of the light distribution characteristics of HID luminaires. New capacitive-switching HID systems that can provide two or more levels of light are discussed in Chapter 12.

Buyer Beware

In some cases, the UL listing of an HID luminaire may be invalidated by HID lamp-ballast *retrofits*. By purchasing *new* HID luminaires, the UL validation issue is avoided.

The selection of the HID luminaire should be based on the following criteria that pertain to the application. These factors are discussed in detail in Chapter 7.

* Color rendering and variation.
* Efficiency.
* Lamp life.
* Lumen maintenance.

- Light distribution.
- Warm-up and restrike time.
- Stroboscopic effects.
- Dimming requirements.
- Visual comfort.

HIGH-BAY FLUORESCENT LUMINAIRES

Fluorescent luminaires have been designed for relatively high mounting heights (up to 30 ft.). Using large, specially designed reflectors, these luminaires typically house T4 quad-tube, or T5 twin-tube, or T5 high-output linear fluorescent lamps.

The unique characteristics of fluorescent operation provide the following advantages over HID systems:

- Instant-on (no warm-up time).
- Instant-restrike.
- Multiple light levels.
- High color rendering.
- High efficacy.

Application Guidelines

Multiple light levels are provided by separately switching each of the 2-lamp or 3-lamp fluorescent ballasts within the luminaire. Using a photosensor, some of the lamps can be automatically turned off to compensate for the available daylight from skylights or windows. This form of "step-level" dimming can be used to provide the appropriate light level in applications such as multipurpose rooms, auditoriums and indoor sports facilities. Three levels of light are usually provided by these systems.

The instant-on and instant-restrike performance allows for automatic on/off control using occupancy sensors. One control option would be to operate the fluorescent luminaires in a bi-level operating scheme. For example, high-bay fluorescent luminaires could be installed to illuminate warehouse aisles that are infrequently occupied. During prolonged periods of non-occupancy in the aisles, these lumi-

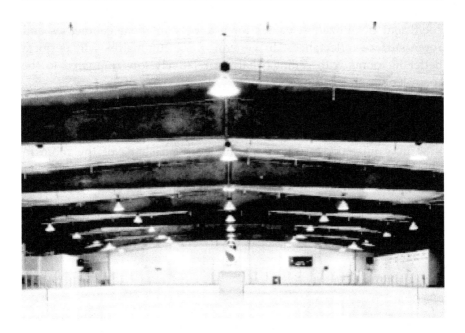

Figure 8-5. High-bay compact fluorescent luminaires are effective in indoor sports arenas, particularly where the ceiling is relatively low compared to the length and width of the room. These luminaires provide good vertical illumination, and with the use of high-frequency electronic ballasts, these light sources do not cause strobe problems. *Courtesy: Sportlite, Inc.*

naires could either be kept off, or held at the 33 percent light level; when the aisle becomes occupied, the occupancy sensor could energize the remaining two ballasts in each luminaire, providing full output without delay.

The diffuse nature of fluorescent lighting improves the illumination of vertical surfaces. This is an important consideration in manufacturing, warehousing, retail and sports lighting.

Buyer Beware

Although fluorescent sources are relatively efficient in terms of lumens per watt, they are not as optically efficient as the HID "point" sources for directing light over long distances. To verify that the high-bay fluorescent luminaires will produce the required footcandles on the floor, ask for the luminaire's photometric data which tabulates the coefficient of utilization values for a variety of room geometries and room surface reflectances. In general, these luminaires will perform better in rooms with ceilings that are relatively low compared to the room length and width.

Have a lighting specifier perform illuminance calculations for specific applications, based on independently measured photometric data. Do not rely on simplified lighting performance tables because they may not take into account the size and shape of the room in which the luminaires are to be located. Alternatively, ask your supplier to arrange either a trial installation or a visit to a similar application where the performance of the system can be directly measured.

Chapter 9

Exit Sign Equipment

E xit signs show the means of egress during an emergency. The routes and the exit doors themselves must be identified with an exit sign. Although the message that these signs communicate is simple, there are many choices of light sources, luminaires and power supply options to consider. This chapter provides an overview of exit sign equipment, emergency lighting regulations, and how to evaluate exit sign performance.

TYPES OF EXIT SIGN LIGHT SOURCES

Although the incandescent lamp has historically dominated the exit sign illumination market, its low efficacy and relatively short life has stimulated the development of alternative light sources. However, carefully evaluate the alternative light sources for their effects on maintained light output, visibility and code compliance before using them in specific applications. Table 9-1 compares the typical performance of exit sign illumination sources.

Long-Life Incandescent

Most incandescent lamps used in exit signs are specifically designed for long operating life. Instead of the nominal 1,000-hour life of standard incandescent light bulbs, exit sign incandescents are rated for 5,000 to over 7,000 hours. One of the most common incandescent exit sign lamps is the 20W T6.5 lamp. Most exit signs utilize two of these lamps operating simultaneously, consuming energy *constantly* at a rate of 40 watts per sign.

Compact Fluorescent

Typically, compact fluorescent exit signs are equipped with two 5W twin-tube lamps, or in some cases, one or two 7W types. Although they are rated to last 10,000 hours, continuous operation usually extends lamp life to approximately twice that of long-life incandescent exit sign lamps, and up to 10 times longer than standard incandescents.

Although compact fluorescent lamps are clearly more efficient than incandescent lamps, there are several other exit sign light sources may provide improved exit sign visibility with far fewer watts. Where codes require that both lamps operate simultaneously, the input to a compact fluorescent exit sign can approach 24 watts.

Table 9-1. Typical performance of exit sign light sources.

Light Source	Watts Per Sign	Life	Rate of Lamp Lumen Depreciation
Long-Life Incandescent	30-40	2-8 months	moderate
Compact Fluorescent	10-24	1-2 years	moderate
Light-Emitting Diode (LED)	1-5	25+ years	moderate
Electroluminescent	<1	8+ years	very high
Radioluminescent (Tritium)	0	10-15 years	high
Photoluminescent	0	n/a	extremely high

**Light-Emitting
Diode (LED)**

LED exit signs can offer a good combination of long life, low wattage and excellent visibility. Improved optical designs have led to the use of higher efficacy and lower first cost.

The LED source can be specified in exit signs for producing red, green and amber light. Red LEDs are the most efficacious LED sources, followed by green and amber.

Under optimal conditions, a single LED may operate continuously for 80 years. However, when used in exit signs, their rated life may be limited to 25 years. The expected life depends on the type of LED, operating temperature and variations in voltage and current. In addition, because

LEDs are circuited in a combination of series and parallel wiring, a single LED failure will cause several other LEDs to extinguish. Over their long life, the light output of LEDs will depreciate by 20 percent or more. Warranties for LED illumination range from 5-25 years.

Electroluminescent (EL)

Another exit sign technology is the electroluminescent (EL) source. This source consists

Figure 9-1. In this application, the exit sign marks where there is a change in direction of the exit route.

of a phosphor-impregnated panel that is sandwiched between two plate electrodes, one that is clear. The phosphor glows when electricity is applied—typically requiring less than one watt. The brightness of the panel depends on the applied voltage and frequency. The color of the light is determined by the phosphors used in the panel. In exit signs, the most common color of light is green, but filters can be used to produce red.

The EL source offers two advantages in addition to energy efficiency. First, it provides extremely uniform illumination across the exit sign letters. And second, it can be used in thin-profile exit signs, satisfying concerns about aesthetics or physical size.

At end of life, EL panels don't extinguish; they just slowly deteriorate in uniformity. And over their life span, the light output diminishes rapidly. Although many EL products are rated for an 8-year life, the typical EL panel will produce only about 20 percent of its initial light output after six years. Some EL products will lose half of their initial light output over the first year of operation. Therefore, its rated life should be adjusted to the period of time that the EL panel is expected to produce the required letter brightness.

Radioluminescent

Radioluminescent tubes produce light without any electrical power. They are filled with a tritium gas that emits beta particles. When the beta particles strike the phosphor coating on the inside of the tubes, light is produced. The color of the light is determined by the phosphors used. In exit signs, the light is usually green.

Because tritium is classified as a low-level nuclear waste, disposal of spent tubes can be problematic. However, manufacturers may recycle the tritium from spent tubes, provided that replacement tubes are purchased as part of the transaction. The tritium poses no health risk while it is contained within the sealed tube. However, if the tube breaks, federal regulations require the evacuation of enclosed spaces near the spill until the tritium can be contained and removed. Tritium is especially dangerous because of its potential for rapid uptake into the body.

Lamp lumen depreciation occurs in tritium tubes as a function of their radioactive decay. Because the half-life of tritium is 12.3 years, the tritium signs will lose half of their light output over this period. Again, it is important to plan to replace the tubes before they are expected to deliver non-compliant exit sign brightness.

Photoluminescent

Another form of self-powered exit signs use photoluminescent—or glow-in-the-dark—materials. These materials absorb ambient light, and they reradiate the stored light when the lights go out. However, because the sign's brightness drops by as much as 90 percent *during the first hour,* they seldom meet building code standards for exit signs. However, they may be used to supplement emergency egress systems that do meet code requirements.

TYPES OF EXIT SIGN LUMINAIRES

The wide variety of exit sign luminaires delivers a range of efficacy, visibility and aesthetics. Note that not all sources can be used in all luminaire types. Table 9-2 indicates the light sources that can be used in each exit sign type.

Table 9-2. Compatibility of exit sign sources and types of signs.

	Incandescent	Compact Fluorescent (CFL)	Light-Emitting Diode (LED)	Electroluminescent (EL)	Radioluminescent
Panel	•	•			
Stencil	•	•	•	•	•
Edge-Lit	•	•	•		
Matrix			•		

Panel

The face of a panel exit sign consists of a single translucent panel. Both the letters and the background are illuminated. For example, an exit sign with red letters and a white background would be a panel exit sign.

The light sources used in panel exit signs must be white. If a red LED retrofit is used, the white part of the exit sign will appear pink, thereby reducing contrast and visibility. Existing panel exit signs are

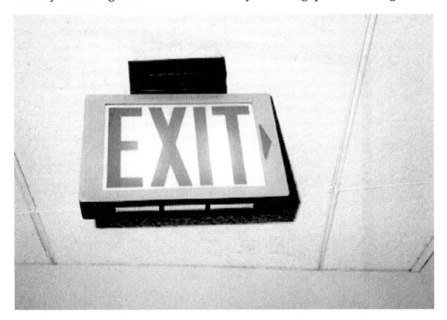

Figure 9-2. A panel exit sign is easily identified by its luminous panel; both the letters and the background are illuminated. Only white light sources should be used in these types of exit signs.

good candidates for retrofitting with low-wattage incandescent or compact fluorescent technologies.

Stencil

In a stencil exit sign, only the letters are luminous; the background is opaque. Unlike the panel exit sign, any of the illumination sources described above can be used in a stencil exit sign. For upgrading stencil exit signs, consider maximizing energy savings with light-emitting diode or electroluminescent retrofit kits.

Edge-Lit

Edge-lit exit signs provide superior uniformity. In these signs, the light is distributed from a sealed chamber through a transparent plate that has the letters etched in or attached to its surface. The sign face appears luminous as light leaves the plate.

Matrix

Matrix exit signs are almost exclusively illuminated with LEDs. Indi-

Figure 9-3. Stencil exit signs have an opaque background. Both white and colored light sources can be used in stencil exit signs. *Courtesy: National Lighting Bureau.*

vidual bare-bulb LEDs form the letters against an opaque background. Because LEDs tend to be directional in light output, matrix exit signs provide excellent visibility when viewed straight on, but their visibility declines at oblique angles. Strings of LED outages are especially apparent in matrix exit signs. When an optical diffuser and a stencil face are added to the design, the letter brightness becomes more uniform. Because most matrix exit signs are LED, they are not candidates for retrofit.

REGULATIONS

The Life Safety Code

Most local building codes are based on the national model: Section 101 of the National Fire Protection Association's "Code for Safety to Life from Fire in Buildings And Structures," commonly referred to as the Life Safety Code. With regard to exit signs, the Life Safety Code's requirements include:

Figure 9-4. Edge-lit exit signs are clear panels with the light source(s) mounted across the top edge. As light passes down through the panel, the letters are illuminated. *Courtesy: Lithonia Lighting.*

Exit Sign Location

No part of an exit route should be more than 100 ft. from the nearest visible exit sign.

Continuous Operation

All signs are to be constantly illuminated, though they may flash upon activation of an alarm system.

Operation During Power Failure

The escape-route emergency lighting must be maintained for 90 minutes after a power failure.

Exit Sign Lettering

The letters must be plainly legible, at least 6 inches high and have a stroke width of at least 3/4 inch.

Brightness

Internally illuminated signs must match the visibility of an externally illuminated sign with a surface illuminance of 5 footcandles. Exceptions are granted for electroluminescent and radioluminescent exit signs because of their high uniformity of letter brightness—these signs are allowed to pass with a minimum brightness of 0.6 footlamberts.

Contrast Ratio

The contrast ratio between the letters and the background must not be less than 0.5, on a scale of zero to one. The closer the contrast ratio is to one, the more visible the letters are against the rest of the sign face.

UL Compliance

Many utilities and building code enforcement officials require that exit signs and/or retrofits comply with the standards set by Underwriters Laboratories (UL). UL's standard for emergency lighting is UL 924, which gives more detailed specifications of the construction, visibility and performance of exit signs. Exit signs with a UL "listing" have met the criteria in this standard.

Retrofits can be UL "listed" if UL's tests indicate that they can be used safely in any existing "listed" exit sign. Alternatively, a retrofit kit may be UL "classified" if it can be used only in specific exit sign models. Installing a retrofit kit that is not UL listed, or not UL-classified for the specific sign, may invalidate the UL listing of the sign in which the kit is being installed.

Other Applicable Codes

Other U.S. codes that are used by local jurisdictions in developing their emergency lighting regulations include:

NFPA 70   National Electrical Code (NEC)

NFPA 99   Standard for Healthcare Facilities

NFPA 497M   Standard for Electrical Installation for Hazardous Locations

NFPA 110   Standard for Emergency And Standby Power

NFPA 171   Standard for Fire Safety Symbols

BOCA   Building Officials & Code Administrators, Basic Building Code

UBC   Uniform Building Code, International Conference of Building Officials

Local Building Codes

Local building codes address the qualifications for exit signs used in a building. Jurisdictions (states, cities, counties, or towns) adopt or modify the model codes or develop their own codes.

Some jurisdictions have specific requirements for exit signs that modify, extend or ignore the requirements of the various model building codes. Because these building codes vary dramatically from location to location, it is essential to become familiar with the local jurisdiction's regulations before proceeding with an exit sign upgrade.

TYPES OF EXIT SIGN POWER SUPPLIES

Because most exit signs are required to continue operating in the event of a power failure, each sign will either be connected to a backup power supply or be self-luminous (requiring no power to operate). Before proceeding with exit sign upgrades, determine what kind of backup power supply is provided. The following is a brief overview of the most common options for emergency power to exit signs.

Battery Backup: Individual Inverter Systems

During a power outage, these exit signs are powered by individual batteries with inverters that convert the battery's DC power to AC power at a frequency of 60 Hz or higher. When the power is returned, battery charging resumes. Because some inverters do not accept all lamp types, check with the manufacturer for compatibility of these systems with retrofit products under consideration.

Battery Backup: Central Inverter Systems

Instead of each exit sign having a dedicated battery system, a central inverter system is used to provide AC power to many exit signs. Large battery assemblies are charged when normal AC power is available; in a power failure, this stored electricity is converted to 60-Hz line voltage using an inverter. These systems can power any retrofit exit sign device, as well as fire alarm systems and other crucial 60-Hz equipment.

Battery Backup: DC Systems

These dual circuit systems consist of a "normally on" AC power circuit and a "normally off" DC lamp/battery circuit. Each exit sign contains two line-voltage AC lamps and two low-voltage DC lamps. In the event of a power failure, the "normally off" DC circuit switches on to provide DC power to incandescent or halogen lamps in the exit signs. The battery systems can be either centrally located for serving many exit signs, or they can be unit systems which serve only one exit sign. These exit signs do not pose any compatibility issues when retrofitting the "normally on" illumination sources, except that the DC lamps may obstruct the positioning of some retrofit products.

Central Standby Generator

Upon power failure, an engine generator provides 60-Hz line voltage to the emergency lighting system. Typically used for backup power in large buildings, these systems must meet requirements in the Life Safety Code for the maximum allowable start-up time.

PERFORMANCE CHECKLIST FOR EXIT SIGNS

Generally, there are three primary performance criteria that facility managers use to evaluate an exit sign alternative: It must be code-com-

Figure 9-5. Upon loss of utility power, emergency lighting systems are designed to switch on automatically and illuminate the exit route. *Courtesy: National Lighting Bureau.*

pliant, economical and highly effective in the event of an emergency. Use the following criteria as a checklist when evaluating exit sign options.

Code Compliance

Verify that the new exit sign complies with Underwriters Laboratory Standard UL 924. To maintain the UL listing in retrofitted exit signs, use only *UL-classified* retrofit kits that are designed for the specific exit sign in question. In addition, contact the local building inspection authority regarding the acceptability of alternative exit sign illumination sources.

Life-cycle Economics: Net Present Value

To determine the most financially attractive exit sign upgrade, consider all of the costs that will occur during the life cycle, including installation, energy, maintenance and disposal. Table 9-3 compares new fixture and retrofit options to an incandescent base case. (The exit signs with the highest NPV will be the most economical.)

Reliability

Reliability is of utmost importance for exit signs. One of the least reliable types of exit sign light sources is the incandescent lamp; with its short life, it is more likely to be burned out when needed in an emergency situation. Of all the alternatives, LED exit sign sources have the longest rated life, exceeding 25 years. Radioluminescent and electroluminescent sources also have relatively long life spans.

Visibility

All exit signs must be highly visible both when normal AC power is available and when powered by the backup system. Exit sign visibility is typically evaluated on the basis of luminance (brightness) and luminance ratio (brightness contrast between the letters and the background). Generally, all exit sign technologies—with the exception of self-luminous exit signs—perform comparably in terms of visibility over a wide spectrum of operating conditions. However, radioluminescent and photoluminescent exit signs can be unreadable at 100 ft., particularly in smoke conditions.

Note that the light output of all light sources depreciates over time. Due to the very rapid light output depreciation of electroluminescent, radioluminescent and photoluminescent sources, request information about the lumen depreciation performance of the products under consideration, and determine when it will be necessary to replace the source to maintain acceptable light output. In general, these light sources will need to be replaced long before they fail.

Some emergency lighting systems can perform optional functions to increase visibility when operated on emergency power. To implement these options, the chosen exit sign must be designed to perform the desired functions. These optional functions include:

Increased Sign Luminance

The brightness of electroluminescent and LED exit signs can be increased by the emergency power system in the event of a power failure.

Flashing

Some exit sign sources, such as LED, electroluminescent and incandescent, can flash on and off during a power failure and/or when a fire alarm circuit is activated. Fluorescent lamps and self-luminous sources cannot be flashed.

Table 9-3. Exit sign upgrade economics.

	Annual Energy Cost	Annualized Maintenance Cost	Upgrade Cost	Net Present Value (NPV)
Incandescent (base case)	$28.00	$19.50	n/a	n/a
New CFL Exit Sign	$ 7.00	$ 9.50	$116.00	$296.00
CFL Retrofit	$ 7.00	$ 9.50	$ 30.00	$377.00
New LED Exit Sign	$ 3.50	$ 0	$116.00	$466.00
LED Retrofit	$ 2.80	$ 0	$ 45.00	$540.00
New Electro-luminescent Exit Sign	$ 0.70	$20.50	$190.00	$166.00
New Radio-luminescent Exit Sign (Tritium)	$ 0	$10.50	$247.00	$252.00

Assumptions: One-sided exit sign; $0.08/kWh; $15/hour for labor; maintenance cost includes lamps and labor for spot relamping upon lamp failure; upgrade cost includes materials and labor; financial analysis based on 20-year cash flow with 3% inflation and 12% discount rate.

Note: Material, labor, energy costs and lamp performance can vary. Contact local suppliers for specific prices and performance data.

Buzzing

Often accompanying a flashing signal, some exit signs will also provide an auditory signal in the event of a power failure and/or when a fire alarm circuit is activated.

Compatibility

There are only a few compatibility issues to consider when selecting retrofit products for upgrading exit signs. These include:

LED Retrofits

LED retrofit lamps and kits should not be used in panel-type exit signs, because their red (or green) color will distort the true color of the panel exit sign face, causing a reduction in letter contrast (visibility). Use only white sources in panel exit signs.

Flashing Sources

Do not use compact fluorescent lamps in an exit sign designed to flash in the event of power failure or fire alarm.

Exit Signs With Rechargeable Batteries

Because the inverter systems used with exit sign batteries are not compatible with all lamp types, check with the manufacturer to determine whether the exit sign light source in question is compatible with the battery back-up system.

Table 9-4. Typical power factor values for exit sign light sources.

Exit Sign Light Source	Typical Power Factor
Compact Fluorescent (CFL)	0.4 - 0.9
Light-Emitting Diode (LED)	0.2 - 0.9
Electroluminescent (EL)	0.2 - 0.8

Power Factor

Switching from conventional incandescent exit signs (which operate at a power factor of 1.0) to more efficient sources with a lower power factor could cause one or more hazards.

The use of exit signs with low power factor can affect the operation and control of a central emergency power system. When standby generators or centralized battery/inverter systems are in emergency operation, the exit signs can form a significant part of the load on the building's emergency electrical system. The electrical distortion caused by a low power factor source could affect the control of that system. As shown in Table 9-4, there are wide ranges of power factor among manufacturers of exit sign upgrades.

Also consider the influence of the power factor on the current-carrying capacity of the emergency power supply and its associated conductors. For a given active power (in watts), an exit sign with a lower power factor will draw more current than a same-wattage exit sign with a high power factor. When upgrading an existing exit sign, verify that the replacement exit sign or retrofit kit does not cause the current to exceed the current-carrying capacity of the existing emergency power supply and its conductors. Excess current can cause wires to overheat (which is a fire hazard), or it can affect circuit breakers or controls.

Chapter 10

Exit Sign
Upgrade Options

E xit signs illuminated with incandescent lamps are easy targets for energy-efficient upgrades that yield rapid returns on investment. Long-life, low-wattage exit sign technologies provide significant opportunities to minimize life-cycle costs by lowering maintenance and operating costs by over 90 percent. In this chapter, we will review popular exit sign upgrade options.

EXIT SIGN UPGRADES

Compared with other lighting system upgrades, exit sign upgrades require relatively little planning and capital. Occupant disruption and acceptance are issues that rarely need to be addressed. And investments in exit sign upgrades carry little risk; they will continue to operate and save energy 24 hours per day, 8,760 hours per year.

Common to all exit sign retrofit kits are adapters that simply screw into the existing incandescent sockets. To avoid inadvertent relamping with incandescents, retrofit kits are available for hard-wire installation. Whatever connection methods are used, installation is relatively easy, usually taking 15 minutes or less per sign. In order of decreasing wattage, exit sign upgrade options are presented in Table 10-1 and in the text that follows. This chapter concludes with a brief discussion of alternatives to consider for new exit signs.

Compact Fluorescent Retrofit Kits
Compact fluorescent retrofit kits are an assembly—consisting of a ballast, socket and lamp—that rests on the bottom of the exit sign enclosure. Where two-lamp operation is required, the assembly may include two ballasts, one for each lamp (consuming up to 24 watts). Some 2-lamp

161

kits utilize a single 2-lamp compact fluorescent ballast for reduced cost and improved efficacy.

Application Guidelines

Compact fluorescent retrofits can be economical, primarily because they are among the least expensive retrofit options. In addition, they provide relatively high brightness, particularly in panel-type exit signs. Electronic compact fluorescent ballasts with high power factor and low harmonic distortion are available for exit sign retrofits.

Buyer Beware

Although CFLs have been recommended for years as an energy-efficient retrofit for exit signs, new technologies—such as low-wattage LED—may provide superior performance in lamp life and energy efficiency.

Table 10-1. Exit sign upgrade options. *Source: EPA Green Lights.*

Source	Retrofit	New Exit Signs
Compact Fluorescent (CFL)	•	•
Light Emitting Diode (LED)	•	•
Electroluminescent (EL)	•	•
Radioluminescent		•
Photoluminescent		•

CFLs cannot be used in exit signs that flash during a power failure or fire alarm. In addition, some exit signs with built-in step-down transformers may not deliver sufficient voltage to operate CFLs.

Although compact fluorescent sources provide excellent brightness levels, this brightness is not evenly distributed across the exit sign face, resulting in reduced uniformity. Trial installations are recommended for verifying that the compact fluorescent retrofit assembly will physically fit inside the exit sign enclosure and that sign

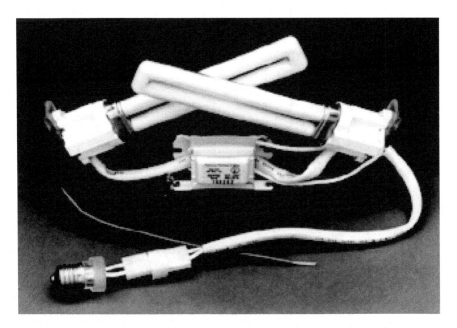

Figure 10-1. This compact fluorescent exit sign retrofit is designed for easy installation in an exit sign with an intermediate incandescent socket. Other interchangeable adapters can be used for candelabra and double-contact bayonet sockets. *Courtesy: ProLight.*

brightness and downlighting (where required) comply with local regulations.

Light-emitting Diode (LED) Retrofit Products

LED retrofit products typically draw only 2-5 watts per sign. Combined with the extremely long rated life of LED sources, this option is one of the most economical retrofits based on life-cycle cost.

One version of the LED retrofit consists of a pair of LED light bars that adhere to the interior of the exit sign enclosure. In some cases, a reflective film may be needed to direct the LED output to the face of the sign. Another version consists of an LED panel that is placed in the exit sign enclosure, suitable for both one- and two-sided exit signs. In addition, simple screw-in LED "lamps" are available, consisting of a series of LEDs encased in a glass or plastic housing.

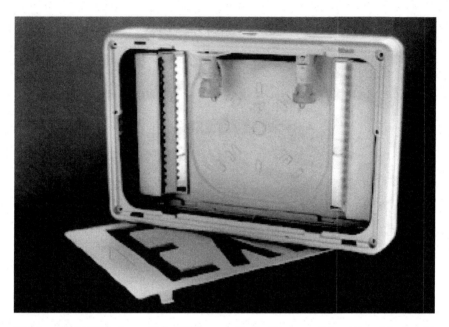

Figure 10-2. This long-life LED retrofit is designed for one-sided exit signs. Each LED light bar receives power through the screw-in adapters. The integral mirrors on each light bar direct the light toward the face of the exit sign. *Courtesy: ProLight.*

Application Guidelines

Some retrofit LED products are specifically designed for either one-face or two-face exit sign applications. With screw-in LED lamps, the brightness of the exit sign can vary depending on the rotational positions of the lamps in their sockets. Conduct a trial installation to confirm acceptable performance.

Buyer Beware

Because of their red (or green) color, LED retrofits can only be installed in stencil-type exit signs. If used in panel exit signs, their color will show through the translucent panel and will result in a discolored appearance, reducing visibility.

Although some building codes require green sources, the efficacy of green LEDs is less than that of red LEDs. Note that the rated life of green LEDs can be reduced if they are operated in a higher current condition to produce code-compliant brightness levels.

Figure 10-3. These one-watt LED retrofit lamps directly replace incandescent lamps in stencil exit signs, reducing exit sign energy costs by up to 95 percent. *Courtesy: Lithonia Lighting.*

Electroluminescent (EL) Retrofit Panels

The lowest-wattage retrofit option is the less-than-one-watt electroluminescent (EL) conversion kit. This retrofit requires the installation of a retrofit EL panel, which has a rated life of at least eight years.

Application Guidelines

Because they are so thin, EL retrofit sign assemblies will fit to very slim-profile existing signs. These kits are available with a battery-powered mini-inverter for battery-backup operation.

Buyer Beware

EL retrofit kits cannot be used for upgrading exit signs that require downlighting capability.

Due to the very rapid lumen depreciation of EL panels, ask the supplier for a warranty that assures compliant levels of brightness over the warranty term.

NEW EXIT SIGNS

There are several situations where it would be prudent to replace exit signs with new models. Some existing exit signs may be in such poor condition or so constricted that it is more cost-effective to replace the sign. For example, exit sign wiring and other fittings may become brittle over time and may fail during a retrofit installation. In addition, exit signs with backup DC-powered lamps have limited space for installing retrofit hardware.

Several choices exist for purchasing new exit signs with consumption of less than 5 watts. Among these choices, radioluminescent sources are the most energy efficient, consuming no electricity. Note, however, that the spent tritium tubes must be disposed of as a radioactive waste. Other new exit sign fixture choices include LED, electroluminescent, compact fluorescent and photoluminescent. The most variety of choices is found in the LED exit sign category. New LED exit signs can be purchased in matrix, edge-lit or stencil configurations.

When considering the purchase of new exit signs, refer to the performance criteria discussed in Chapter 9.

Chapter 11

Switching Controls

R egardless of how efficient a lighting system may be, energy dollars are wasted whenever the lights are left on unnecessarily. Properly installed, automatic switching controls will eliminate this wasted expense. Combined with luminaire efficiency upgrades, automatic switching controls are an essential ingredient for maximizing energy savings and profit.

OVERVIEW OF AUTOMATIC SWITCHING STRATEGIES

This chapter identifies opportunities for automatically switching lighting systems based on occupancy, time or daylight. The selection of the switching strategy will be influenced by the predictability of occupancy, the physical layout of the lighting system, whether the lighting system is located indoors or outdoors, and electricity rate structures.

Occupancy-Based Switching

How many times have the lights been left on when "nobody's home?" Unfortunately, it happens far too often, resulting in costly energy waste. Occupancy sensors can minimize lighting waste by ensuring that lighting operation is limited to times when the illuminated space is actually occupied. The most effective applications are in spaces with *unpredictable* occupancy schedules. Anecdotes about the failures of occupancy sensors to deliver reliable service can usually be traced to human error—not sensor malfunction. The key to successful occupancy sensor installations is the proper specification, location and adjustment of each occupancy sensor. Because each unique space type requires careful analysis for determining correct applications, occupancy sensor projects will require more time for developing the final specification compared with common lighting upgrade projects. In many applications, however, the energy savings potential of occupancy sensors can justify this additional effort in survey and analysis.

Time-Based Switching

In facilities with relatively *predictable* occupancy patterns, timing devices can be used to automatically control lighting operation on a predetermined schedule. User overrides can provide flexibility to accommodate variable occupancy patterns. A vast array of product choices awaits the facility manager who is interested in eliminating unnecessary overnight lighting operation. These choices include timer switches, electronic time clocks and centralized lighting control systems. Beyond simple switching strategies, many of the advanced systems also provide dimming capabilities to increase energy savings and meet load management objectives.

Daylight-Based Switching

Photocells have been used for decades to limit the operation of outdoor lighting systems to nighttime hours. New advances in technology now provide greater accuracy and flexibility for controlling outdoor lighting systems. Although cost-effective daylight-switching opportunities exist in certain *indoor* applications, proceed with caution as occupants may be distracted if lighting in their areas are automatically turned off (or prevented from turning on) when target daylight levels are attained.

OCCUPANCY SENSORS

Occupancy sensors save energy by automatically turning off lights in unoccupied spaces. When motion is detected, the sensor activates a control device that turns on the luminaires. If no motion is detected within a specified period of time, the sensor turns off the lights until motion is sensed again.

Compared with other control options, occupancy sensors provide the *maximum reductions* in lighting system operating hours, particularly where the pattern of occupancy is intermittent or unpredictable. The sensors will not only prevent wasteful overnight lighting operation, but they will also eliminate wasted lighting operation during normal business hours in spaces that are temporarily unoccupied.

Occupancy sensors can be used in a very wide range of lighting control applications and should be considered in every upgrade decision. Occupancy sensors may be installed to provide on/off control of incandescent or fluorescent loads as well as bi-level control of capacitive-switching HID luminaires (that idle in a low-output mode during periods of unoccupancy). Refer to Chapter 12 for a complete discussion of capacitive switching HID luminaires.

Control Adjustments

Most occupancy sensors have adjustable settings for both sensitivity and time delay. Once installed and properly calibrated, occupancy sensors seldom require subsequent adjustments. However, after completing renovations or rearranging furniture, the occupancy sensors may need to be "tuned" by adjusting their sensitivity and/or time delay settings.

Sensitivity

The sensitivity setting allows the user to adapt the sensor for the magnitude of motion that is expected to occur in the space. A proper sensitivity setting will ensure that normal motion is detected without triggering responses to extraneous signals. If the sensitivity setting is too high, the lights may continue to operate in unoccupied spaces; if the setting is too low, the lights may turn off in spaces where normal motion is occurring.

Time Delay

The time delay setting refers to the amount of time that elapses with no motion detected before the luminaires are turned off. The time delay prevents the luminaires from switching off during intervals when people are actually in the room, but move too little or too slowly to be detected by the sensor. A good initial choice for the time delay is about 10 minutes. Studies have shown that when occupants leave a room, they most commonly return *within* two minutes or *after* 15-20 minutes.

Types of Motion/Occupancy Sensing Technologies

The two most common motion- or occupancy-sensing technologies used in occupancy sensors are passive infrared and ultrasonic. Either

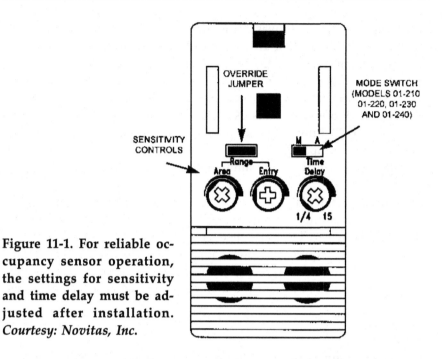

Figure 11-1. For reliable occupancy sensor operation, the settings for sensitivity and time delay must be adjusted after installation. *Courtesy: Novitas, Inc.*

technology can be housed in ceiling-mounted or switch-mounted sensors. Because of the unique characteristics of each sensing technology, there are applications where one technology will provide much more reliable performance than the other. Therefore, a thorough understanding of the differences in motion sensing characteristics is required before occupancy sensors can be correctly specified. In addition, some sensors combine passive infrared with audible-noise-sensing technology for increased sensitivity in areas with partitions.

Passive Infrared Occupancy Sensors
        Passive infrared (PIR) sensors respond to motion between horizontal and vertical cones of detection defined by the faceted lens surrounding the sensor. As an occupant moves a hand, arm or their body from one cone of detection to another, a positive "occupancy" signal is generated and sent to the controller. Because these cones of detection radiate from the sensor, a greater range of motion is required at a greater distance in order for the sensor to detect motion. And because gaps exist between adjacent cones of vision, these dead spots get wider with distance. Note that the PIR sensors require an unobstructed view of the motion and are

much more sensitive to motion occurring lateral to the sensor. Because infrared sensors require direct line-of-sight to the moving object, they will not perform properly in spaces where furniture, partitions, or other objects are between the sensor and the occupant.

Ultrasonic Occupancy Sensors

Ultrasonic sensors are *not* passive; they emit and receive high-frequency sound waves in the range of 25-40 kHz, well above the range of human hearing. These waves reflect off objects and room surfaces, and the sensor measures the frequency of the waves that return to the receiver. If there is motion within the space, the frequency of the reflected waves will shift slightly; the change is detected by the receiver and the sensor registers a positive occupancy signal. Ultrasonic sensors can detect motion that is hidden from the sensor's view, provided that the space is enclosed with hard surfaces to reflect the waves back to the receiver. Ultrasonic sensors are much more sensitive to movement directly toward or away from the sensor, compared to lateral movements. In general, ultrasonic sensors can cover a larger area and are more sen-

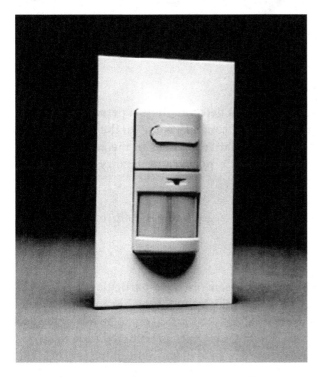

Figure 11-2. Switch-mounted infrared occupancy sensor with manual-on option. This sensor can be configured for automatic-on operation if desired. *Courtesy: The Watt Stopper, Inc.*

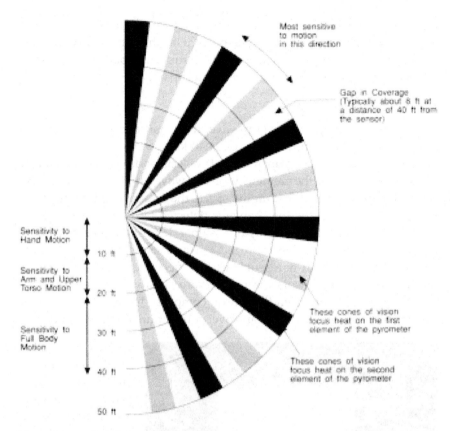

Figure 11-3. For infrared sensors to detect motion, a person must move all or part of their body from one "cone of vision" to the next, crossing a gap between adjacent cones. Relative changes in the amount of heat received in each cone will trigger a positive occupancy signal. *Courtesy: California Energy Commission.*

sitive than infrared sensors. These advantages help to justify their slightly higher cost.

Dual-technology Sensors

Both infrared and ultrasonic sensors have particular strengths in specific applications. And when misapplied, they have unique weaknesses. For example, infrared sensors are most likely to be misapplied so that the lights are turned off while the space is occupied (due to an obstructed view of the motion or excessive distance from the sensor). The

Figure 11-4. Switch-mounted ultrasonic sensor with manual-on option. This sensor can be configured for automatic-on operation if desired. *Courtesy: Novitas, Inc.*

most common misapplication of ultrasonic sensors is that they might keep the lights on in unoccupied spaces due to oversensitivity to extraneous signals (such as air movement or vibrations). Although there are many applications where one or the other technology will perform reliably, there may be situations where maximum reliability is needed. These situations are where dual-technology sensors should be considered.

Although other operation-logic settings can be configured by the user, dual-technology occupancy sensors are usually configured to operate in the following manner:

- Both infrared *and* ultrasonic occupancy signals are needed to *turn on* the lights.

- Absence of both infrared *and* ultrasonic occupancy signals are needed to *turn off* the lights.

- Either infrared *or* ultrasonic occupancy signals are needed to *keep the lights on.*

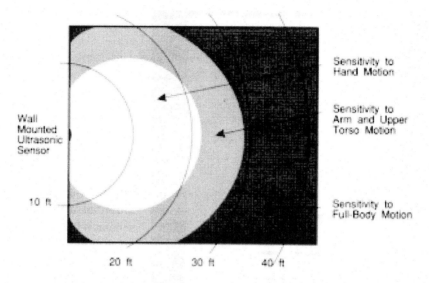

Figure 11-5. Typical sensitivity characteristics of ultrasonic occupancy sensors. *Courtesy: California Energy Commission.*

Figure 11-6. Dual-technology sensors utilize both infrared and ultrasonic technologies for increased reliability. *Courtesy: The Watt Stopper, Inc.*

The combination of both passive infrared and ultrasonic motion-sensing technologies allows the sensor to take advantage of the best features of both technologies while counteracting their weaknesses. As a result, reliability can be increased while improving sensitivity and coverage. Because they can be applied in applications that favor *either* infrared or ultrasonic technology, a single model can be specified in a variety of applications. This feature can contribute to less-diverse occupancy sensor specifications and higher purchasing volumes, which can help offset the higher cost of these units. They are best applied in large-area locations that are intermittently occupied, such as classrooms, large conference rooms and lunchrooms.

Sensor Mounting Locations

Occupancy sensors are available in both ceiling-mounted and switch-mounted versions, utilizing either infrared or ultrasonic sensing technologies. In addition, workstation occupancy sensors can be used for automatically controlling workstation loads such as task lights, computer monitors, printers and radios.

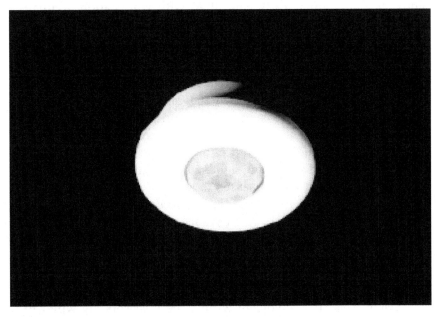

Figure 11-7. Ceiling-mounted infrared occupancy sensor. *Courtesy: The Watt Stopper, Inc.*

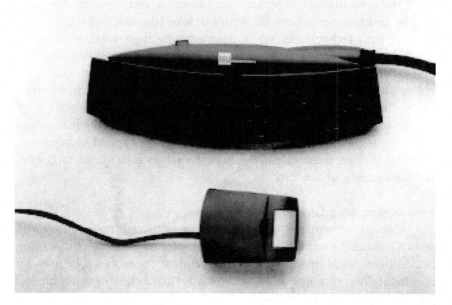

Figure 11-8. Loads such as task lights, heaters, computer screens and radios can be automatically controlled with workstation occupancy sensors. *Courtesy: The Watt Stopper, Inc.*

Switch-Mounted Sensors

Wall-switch occupancy sensors are the least expensive and easiest to install. Typically costing less than $60, installation can be as easy as replacing a standard wall switch. Common applications for switch-mounted sensors include separately switched areas such as conference rooms, classrooms, individual offices and storage rooms. Because these devices are mounted in existing light switch locations, check the coverage pattern provided by the sensor to see if it will adequately detect motion throughout the room. In addition, verify that the *type* of motion in the space will be detected, given the sensor type and its location relative to the dominant motion in the space (see discussion of infrared and ultrasonic technologies above).

Ceiling-Mounted Sensors

Using Class II low-voltage wiring, these sensors are wired to a separate low-voltage power supply and a relay that performs the actual switching function in the ceiling plenum. In large-area applications, multiple sensors can be used with one power supply and relay, but manufacturers specify a maximum distance between the sensors and the power supply for reliable operation. Although the installation costs for ceiling sensors can be more than 2-3 times as much as switch-mounted sensors, ceiling-sensor installations can be cost effective if each sensor controls a relatively large load. Ceiling-mounted sensors should be used in areas where switch-mounted sensors would be inadequate, such as in corridors, open office areas, warehouse aisles and spaces where objects obstruct the coverage of a switch-mounted sensor.

Workstation Occupancy Sensors

Although most occupancy sensors are designed to control the ambient lighting, workstation occupancy sensors have been developed to automatically control workstation "plug loads," such as computer monitors, task lights, space heaters and radios. Typically mounted under desk surfaces or shelves, a passive infrared sensor is used to sense motion in the vicinity of a working area. The sensor is electrically connected to a power strip which disconnects "controlled" loads when the workstation area is unoccupied. "Uncontrolled" loads—such as computer CPUs, fax machines and modems—are plugged into "uncontrolled" receptacles that deliver constant power, regardless of occupancy. The cost-effectiveness of these devices depends heavily on the total wattage of the controllable loads and the local electricity rate.

Occupancy Sensor Control Options

Specifiers of switch-mounted or ceiling-mounted sensors can choose among several unique control options. These options are described below.

Automatic-On

The most common control option specified in occupancy sensor applications is the automatic-on type. Upon entering an unoccupied space, the sensor will automatically turn the lights on. After vacating the space, the sensor will automatically turn the lights off after the preset time delay. Although this operating mode may be the most convenient, more energy can be saved by using manual-on occupancy sensors, as described below.

Manual-On Switch-Mounted Sensors

Manual-on occupancy sensor installations require the occupant to manually energize the lighting system by activating the sensor's on/off switch upon entering the room. Compared to purely automatic-on sensors, this control strategy can save additional energy dollars. For example, when the room has sufficient daylight, the occupant can choose to use the space without activating the lighting system. Of course, these sensors will automatically turn off the luminaires a few minutes after motion is no longer detected. Some manual-on switch-mounted occupancy sensors produce an audible signal just before the lights turn off, in case an occupant is still in the space.

Manual-On Ceiling-Mounted Sensors

When ceiling-mounted sensors supplement the use of the wall switch, the system can operate in either a manual-on or automatic-on mode, depending on whether the last occupant to leave the space manually turned off the lights. For example, if the last occupant manually turned off the lights, the system requires a manual-on operation for the next occupant; if the last occupant left the lights turned on, the system shifts to the automatic-on mode for the next occupant. To ensure that the ceiling-mounted occupancy sensor installation *always* operates in the manual-on mode, install *latching switches* in place of standard toggle switches. Here is how one product, the Sentry Switch, operates: Ten seconds after the ceiling-mounted occupancy sensor turns the lights off, the latching toggle switch will automatically unlatch and drop to the off position, requiring the next occupant to manually activate the lighting system upon entering the room. The 10-second delay is provided just in case any occupants are still in the room when the lights are shut off, at which time they can initiate "major motion" to keep the latching switches from turning off. In these cases, the occupants won't have to physically switch the lights back on.

Bi-Level Switch-Mounted Sensors

Separately switching individual lamps in a fixture is a cost-effective way of controlling light levels. "Dual switching" is encouraged by most of today's building energy codes, and features a double-gang wall box with two switches. Each switch controls the power supplied to one of the two ballasts in each fixture. (Alternatively, the switches may control whether a bi-level electronic ballast produces partial or full output.) Some switch-mounted occupancy sensors have been specifically designed to maintain this dual-switching, bi-level flexibility. The user can select whether the system will provide full or partial light output when occupancy is sensed. In three-lamp dual-switching applications, the partial lighting choice can be wired for either one-lamp *or* two-lamp operation.

Daylight Switching Occupancy Sensors

These sensors add to the savings normally expected from occupancy sensors. Available in both ceiling and switch-mounted versions, a daylight sensor is included that can be calibrated to turn off the lights (and/or prevent lights from turning on) when ambient light levels reach a desired target. Users will find that the *ceiling-mounted* daylight sensing controls are more accurate in measuring task luminance compared with switch-mounted daylight/occupancy sensors. This limitation may restrict the use of these switch-mounted units to areas where light level measurement is not critical, such as in lunch rooms. Note that because daylight *dimming* is not as conspicuous to occupants, daylight dimming is usually preferred over daylight switching in indoor applications. A trial installation is recommended to assess user acceptance of this technology.

Evaluating Occupancy Sensors

Not all sensors perform comparably. Before purchasing a specific name-brand sensor, conduct a simple trial installation of the various products under consideration. Follow the procedure below for conducting a test:

1. Install the sensors temporarily in a strategic location as suggested by the sensing coverage pattern.

2. Connect these sensors to a power supply. They do not, however, need to control the lighting circuit.

3. Notice the LED indicator light that illuminates when the sensor detects motion. At various locations in the test room, perform several types of motions, varying the magnitude, speed and direction of motion. Also, include a test that evaluates the sensor's ability to detect motion behind obstacles.

4. Note which sensors were most successful in detecting minor motion (both with and without obstacles), as well as which sensors

Saving Energy Via Wattage *and* Time Reduction

By combining watt-reducing upgrades and hour-reducing upgrades into a single project, energy savings can be maximized—with typical savings of over 60 percent—while yielding highly profitable returns. The example below (source: EPA Green Lights) presents the results of three different upgrades proposed for a manually switched 1,000-fixture system. The base luminaire uses four F40 fluorescent lamps and two standard magnetic ballasts. The three upgrades are:

1. A watt-reducing upgrade to T8 lamps and partial-output electronic ballasts (only).

2. An hour-reducing installation of occupancy sensors (only).

3. A simultaneous upgrade of T8 lamps with electronic ballasts and occupancy sensors.

This example illustrates how utilizing a comprehensive approach which includes both lighting upgrades and switching controls (Option #3) can result in maximizing energy savings and producing the highest net profit. Although Options #1 and #2 yield shorter payback periods, these options are less profitable because of their lower net present value.

	OPTION #1 T8 Lamps & Electronic Ballasts (Only)	OPTION #2 Occupancy Sensors (Only)	OPTION #3 T8 Lamps, Electronic Ballasts & Occu- pancy Sensors
Percent Energy Savings	50%	30%	65%
Project Cost	$ 58,000	$30,000	$ 88,000
Internal Rate of Return (10 yr.)	57%	57%	45%
Net Present Value (10 yr. @ 12% dis- count rate)	$143,900	$76,400	$158,500

Conservative assumptions were used for the costs of materials, labor and maintenance. Annual hours of operation were assumed at 3,500, with the occupancy sensors saving 30%, controlling an average of 3 luminaires per sensor and costing an average of $90.00 per sensor, installed.

were most affected by false signals (such as sensing motion in adjacent corridors).

Application Guidelines

Occupancy sensors—when properly specified, installed and adjusted—should provide reliable operation of lighting systems during periods of occupancy and should not disrupt normal business activity. Most causes of failed occupancy sensor installations can be linked to improper product selection and placement. By following the guidelines below, occupancy sensor installations should provide significant energy savings.

Occupancy Patterns

The most favorable applications of occupancy sensors are in spaces that are intermittently (or unpredictably) occupied. The first places to look for cost-effective applications of occupancy sensors are in common or public areas—such as rest rooms, conference rooms, storage areas, printer/copier areas, snack areas and corridors—where occupants expect that others are responsible for controlling the lighting operation. Operating-hour reductions in these areas can yield energy savings of 30 to 75 percent. Other occupancy sensor applications—such as private offices, open office areas and warehouse aisles—can be cost-effective depending on the occupancy patterns and manual switching practices. For example, some office building cleaning personnel are instructed to turn on all the lights in the building or floor, and turn them off as each space is cleaned. The installation of occupancy sensors in these applications can provide more savings than in buildings with other cleaning management practices.

Infrared Sensor Guidelines

Follow the guidelines below to maximize the success of infrared sensor installations.

• Line-of-sight is required between the sensor and the occupant motion. Infrared sensors may not work well where partitions block direct viewing of occupants.

• The magnitude of motion required to keep the lights on is directly proportional to distance between the occupant and the sensor.

Sensor Technology	Private Office	Large Open Office Plan	Partitioned Office Plan	Conference Room	Rest Room	Closets/ Copy Rooms	Hallways Corridors	Warehouse Aisle Areas
US Wall Switch	●			●	●	●		
US Ceiling Mount	●	●	●	●	●	●		
IR Wall Switch	●			●		●		
IR Ceiling Mount	●	●	●	●		●		
US Narrow View							●	
IR High Mount Narrow View							●	●
Corner Mount Wide View Technology Type		●		●				

Figure 11-9. Use this chart as a guide when selecting occupancy sensor products. Always request supplier assistance when specifying and positioning occupancy sensor products. *Courtesy: EPA Green Lights.*

Many infrared sensor applications fail when the wall-switch sensor is located at the opposite end of a long room from where the motion typically occurs.

- Infrared sensors are least sensitive to motion toward and away from the sensor. They are most sensitive to motion lateral to the sensor.

- Infrared sensors work well outdoors and in high-bay areas. Their applications do not need to be restricted to enclosed spaces. In high-mount applications, infrared sensors are effective in restricting occupant-sensing coverage to specific areas (such as warehouse aisles).

Ultrasonic Sensor Guidelines

In many applications, ultrasonic sensors perform much differently than infrared sensors. The application guidelines below highlight these differences.

- When used in enclosed spaces with hard surfaces, ultrasonic sensors do not require direct line-of-sight to the occupant. However, in open office areas, particularly those with fabric-covered partitions, ultrasonic sensors may require line-of-sight for reliable motion sensing.

- The magnitude of required motion increases with distance between the ultrasonic sensor and the occupant. Verify that the sensor's coverage pattern exceeds the size of the space where motion is expected to occur.

- Ultrasonic sensors are least sensitive to motion lateral to the sensor. They are most sensitive to motion toward and away from the sensor.

- In high-bay warehouse applications, ultrasonic sensors are generally unsuitable for restricting the area of occupancy coverage to specific aisles. The ultrasonic sound waves can wander over the shelves into adjacent aisles and reflect back to the receiver, causing false triggers.

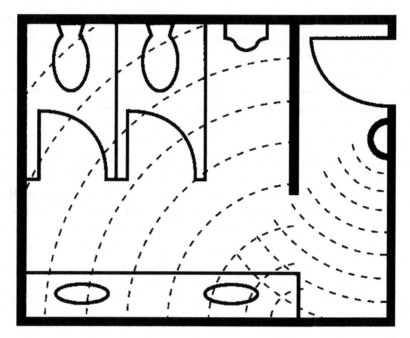

Figure 11-10. Because sound waves can reflect off hard surfaces, ultrasonic occupancy sensors are the best choice for restroom applications. *Courtesy: MyTech Corporation.*

Sensor Coverage

Product performance information provided with the occupancy sensor will indicate the coverage area which defines the physical limits of the sensor's ability to detect motion. Most occupancy sensor manufacturers publish their coverage areas based on the maximum sensitivity setting, although this may not be clearly stated in the product literature. Some of the published coverage patterns indicate the zones where only minor motion (hand movement) and where major motion (full-body motion) is required to keep the lights on.

Sensor Placement

The specification and placement of occupancy sensors should be performed by an experienced professional to ensure adequate occupancy sensing coverage. In large areas, more than one occupancy sensor may be required in a space to extend the coverage area. The maximum horizontal spacing of ceiling-mounted occupancy sensors in large open office areas can be affected by the use of partitions. As the partition height

increases relative to the ceiling height, the sensor spacing must be reduced to provide line-of-sight coverage. Check with the manufacturer
for their recommended spacing of ceiling-mounted sensors when used in
partitioned areas.

Buyer Beware

To gain the financial advantages of an occupancy sensor installation, a careful room-by-room survey must be completed, keeping a
watchful eye for misapplications that could hinder the project's success.

Post-Installation Sensor Adjustments

Occupancy sensor systems must be "tuned" after installation. Most
suppliers offer this post-installation service which involves adjusting
sensitivity and time delay settings as appropriate for the space. As part
of the agreement with the supplier, require a minimum 24-hour response
time to address occupant complaints that may arise after the sensors
have been installed and tuned. In some cases, the placement of sensors
may need to be adjusted. To ensure proper operation, perform the following simple tests while observing the LED light on the sensor that
indicates when motion is detected.

- *Entry Test.* Sensors should activate the lighting within two seconds
  after a person progresses three feet into the space. Verify that the
  sensor does not activate when a person passes outside the room
  with the door open.

- *Hand Motion Test.* Sensors should activate the lighting when hand
  motions of one-foot distance back and forth are made in various
  directions. Note the direction and magnitude of motion required to
  activate the sensor.

- *Perimeter Test.* Perform hand and body movement at various locations around the room to determine areas where the sensor is
  least effective in detecting motion. If the sensitivity setting is increased, verify that the lights turn off at the end of the time-delay
  cycle and that motion outside, room vibrations, or strong air currents do not provide a false signal and keep the lights on. Although increases in time delay will help guard against "false-off"
  occurrences, an excessive time delay setting will unnecessarily reduce energy savings.

False Signals for Ultrasonic Sensors

Ultrasonic sensors can be activated by vibrations (which, for example, may be caused by the starting of an air conditioner). Also, ultrasonic sensors can be activated by moving air and should not be used in areas where strong air currents exist; ceiling-mounted ultrasonic sensors should be located away from ventilation diffusers. In some cases, ceiling-mounted sensors will pick up motion that occurs outside an open doorway. Performing a trial installation and attempting these "false" signals can aid in proper sensor specification and placement where these occurrences are common. Some sensors are designed to filter out false signals caused by repetitive or constant action, such as air flow.

False Signals For Infrared Sensors

Although infrared sensors are quite resistant to false triggering, they may be located in positions that allow the sensor to have line-of-sight into an adjacent corridor which could keep lights on unnecessarily. By applying a masking material to the appropriate facets of the PIR sensor's lens, this potential problem can be avoided. In addition, direct sunlight on the lens may provide a false signal to the PIR sensor.

Compatibility With Electronic Ballasts

Mechanical relays typically used in older-technology occupancy sensors may become damaged by the relatively high inrush currents that result from an occupancy sensor's making and breaking of electrical contact in low-harmonic (<10% THD), electronically ballasted fluorescent systems. Contact the supplier to verify that their occupancy sensors are compatible with electronic ballasts.

Load Limits

Check the manufacturer's literature regarding the maximum (and in some cases, minimum) loads that can be switched. Most switch-mounted sensors can handle up to about 700 watts of fluorescent lighting; ceiling sensors typically switch up to 2,000 watts of fluorescent lighting.

Electricity Rate Structure

The electricity *rate structure* can also have a major influence on the cost-effectiveness of an occupancy sensor installation. When the building's electricity rate does not include a demand charge, the percent savings in operating hours will be comparable to the percent savings in energy costs. However, if demand charges represent a large portion of the electricity bill, the dollar savings from the sensor installation may be considerably less than the energy savings (in percent). This is especially true if much of the energy savings occur during off-peak (overnight) periods when demand and/or energy rates are lowest. To calculate energy cost savings, use weighted average rates for energy and demand charges that represent the costs *avoided* by the installation of occupancy sensors. See Chapter 18 for a complete discussion of the impacts of rate structures on energy cost savings calculations.

Impacts on Fluorescent Lamp Life

It is well known that the frequent switching of fluorescent lamps can reduce lamp life. In addition, the *type of starting* (rapid-start or instant-start) provided by the ballast also has a significant impact on lamp life, with *instant-start* systems having the greatest effect. However, before trading the energy savings of occupancy sensors and instant-start ballasts for lamp-life preservation, consider:

• Lamp replacement costs are directly related to how frequently lamps are replaced, *not* on how many burn hours the lamp delivers. The lamp's *calendar life* is used to determine how many dollars will be spent on replacing failed lamps. In *most* occupancy sensor applications, the percent reduction in expected lamp life (in hours) is exceeded by the percent reduction in the annual operating hours over its life, resulting in a net *increase* in lamp calendar life.

$$\frac{\text{Calendar Lamp}}{\text{Life (years)}} = \frac{\text{Estimated Lamp Life (hours)}}{\text{Annual Hours of Operation (hours/year)}}$$

• Even in the unlikely event that the lamps may need to be replaced more frequently, there are many situations where the added energy savings from the use of instant-start T8 electronic ballasts will outweigh the costs associated with more frequent relamping.

TIMER SWITCHES

Although spring-loaded timer switches have been marketed for decades, *programmable electronic* timer switches have been introduced. These simple devices replace conventional wall switches. After a predetermined time interval following the "switch-on" operation, the timer switch will turn off the lights. Depending on the technology used, the lights-on period can be chosen either by the installer or by the occupant. Upon exiting the space, the occupant can (and should) turn off the lights before the programmed lights-on period has elapsed. These switches can be programmed to provide a warning signal before the lights are turned off.

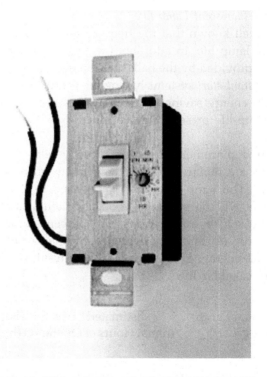

Figure 11-11. This electronic timer switch will automatically turn off power to the circuit after a preset time interval. *Courtesy: Paragon Electric Company, Inc.*

Application Guidelines

Costing as little as $30 each, timer switches can be an economical alternative to occupancy sensors, especially when the duration of occupancy is typically short-term or repeated regularly. Short-term occupancy applications would include self-storage facilities or library stacks. Applications in spaces with regular occupancy hours can include industrial lighting or customer service office lighting where the work is scheduled in shifts.

Buyer Beware

The time interval programmed into the timer switch should be carefully selected. Too long an interval can waste energy; too short an interval can cause occupant frustration and safety concerns, possibly threatening the continued use of the technology.

In short-duration applications, it is possible that as occupants leave, subsequent occupants may not reset the timer after the initial occupants leave, which could leave them in the dark after a few minutes. In most timer-switch applications, the controlled lighting system should be wired to include a few luminaires that are energized on a separate, uncontrolled (always-on) circuit.

Occupancy sensors should be considered as a competing technology for timer-switch applications. The additional cost of occupancy sensors may be justified by their superior energy savings performance and occupant convenience.

ELECTRONIC TIME CLOCKS

Electronic time clocks are fully automatic load switching systems that turn lighting systems (and other loads) on and off according to a preset schedule. Electronic, programmable, microprocessor-based time clocks are more expensive than the older-technology mechanical time clocks, but they provide up to 365-day programming flexibility and can be programmed to adjust for daylight-savings time and leap years. And their electronic construction offers more reliable and accurate on/off scheduling. Most electronic models include a backup system for maintaining the programmed schedule in the event of a power failure.

Application Guidelines

Electronic time clocks are generally used to control relatively few circuits simultaneously. Individual operating schedules can be programmed for each circuit. Typical applications include small-business retail lighting and common-area lighting in apartments and small office buildings. Some time clocks are specifically designed to control outdoor lighting loads—such as lighting billboards and parking lots—by making daily adjustments in sunrise and sunset times according to predicable astronomic patterns.

Buyer Beware

Time clocks are programmed to switch loads strictly according the preset operation schedule, which may or may not coincide with occupancy. Consider occupancy sensors for saving additional energy and

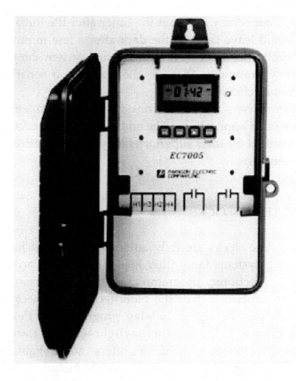

Figure 11-12. This four-channel electronic time clock can store switching schedules for each of four circuits. *Courtesy: Paragon Electric Company, Inc.*

providing superior flexibility to changing patterns of space use.

Some electronic time clocks deliver a small amount of power to the lamps, even when the lights are programmed to turn off. This electricity "trickle" can shorten the life of preheat CFLs. Check with the manufacturer to determine if a trickle current is used in the electronic time clock.

CENTRALIZED LIGHTING CONTROL SYSTEMS

Usually included as part of a building's energy management system (EMS), a centralized lighting control system provides essentially the same load scheduling functions as an electronic time clock, but on a larger scale. In addition, centralized systems can perform more sophisticated functions, including dimming and load management, if proper software and lighting equipment are installed.

Centralized lighting control systems involve the installation of a relay cabinet near the existing electrical panel. Relays are simple switches

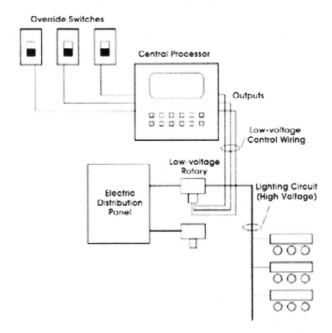

Figure 11-13. Typical installation of a centralized lighting control system. *Courtesy: California Energy Commission.*

that receive low-voltage ON-OFF switching signals from the programmable scheduling microprocessor. A relay is installed in each lighting circuit that leads from the electrical panel. Some manufacturers of low-voltage relay panels include the scheduling microprocessor in the same cabinet with the relays. Low-voltage switches can be installed adjacent to the cabinet or in remote locations to give occupants manual and override control.

Remote switching control is installed by pulling low-voltage wire from the relay panel to each switch location. Alternatively there are two methods for communicating control signals to the lighting control system without pulling new wires through the building. Communicating via existing telephone cables, occupants can perform system overrides in their areas using their touch-tone telephones. Power-line carrier (PLC) systems are also easy to install because the override signal can be transmitted from the override switches to the relays by using the existing power lines. However, PLC systems are highly susceptible to power line interference and may behave erratically when used in conjunction with electronic ballasts or other electronic devices. Check with the suppliers regarding the compatibility of specific electronic ballast models with the PLC communication system.

Application Guidelines

Sweep systems are a common form of centralized switching control. These systems establish a programmed schedule for sequentially turning off lights throughout a floor or an entire building. A typical application is found in office buildings, where the systems ensure that lighting is not unnecessarily left on by the occupants. For example, if most of the occupants on a given floor normally leave by 6:00 PM, then the system will provide a warning signal (such as flicking the lights off and on) a few minutes prior to turning the lights off in the space. This warning signal allows any remaining occupants to override the scheduled lighting "sweep" in their location. This override may need to be repeated periodically until the space is unoccupied.

A less expensive version of sweep systems involves the use of latching switches. When an electronic time clock turns power off to a circuit, the latching switch will unlatch and mechanically return to the off position. After 2-5 seconds, the occupants can find their applicable latching switch (which glows) and turn their local lighting system on again manually. This upgrade requires no additional control wiring and is installed in the same way as a standard wall switch. The primary disadvantages of this technology are that a momentary power interruption will affect all lights controlled by latching switches and that the occupants do not receive a warning signal when the lights are about to be extinguished.

Buyer Beware

Unless an EMS already exists in the building, installing a retrofit central lighting control system may be cost-prohibitive. Most systems are priced on the installed cost per point of control. To maximize cost-effectiveness, the largest allowable lighting load is designed to be switched by a single control point. However, after a comprehensive lighting upgrade, the lighting load may have been reduced to the extent that the system will not be fully utilized (watts per point) to cost-justify the installation.

Unlike occupancy sensors, scheduling systems do not have the flexibility to eliminate wasted energy consumption during normal business hours.

Selected luminaires should operate on a 24-hour circuit in areas with sweep systems to provide safe access to lighting control override switches.

DAYLIGHT SWITCHING SYSTEMS

Photocells or scheduling systems can be used to automatically turn off lighting systems when sufficient daylight is available. Daylight switching upgrades are among the most profitable lighting control upgrades available.

Application Guidelines

All outdoor lighting should be controlled using a daylight switching system. In many cases, photocells have been used to automatically provide "dusk-to-dawn" operation. The resulting operating hours under photocell control is typically 4,100 hours per year, because the lights are typically turned on about 20 minutes after sundown and turned off about 20 minutes prior to sunrise.

In applications where the outdoor lighting is not needed for dusk-to-dawn illumination, a timed switching system may be wired in series with the photosensor to switch off the circuit before dawn. For example, a retail establishment may require high-level parking lot illumination

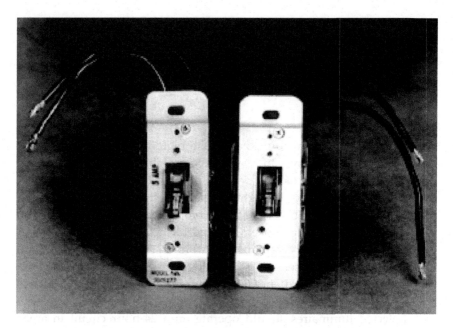

Figure 11-14. Latching switches can be used with electronic time clock controls to provide an inexpensive means for user override control. In addition, latching switches can be used with ceiling-mounted occupancy sensors for manual-on operation. *Courtesy: The Watt Stopper, Inc.*

from dusk until one hour after closing—say 11:00 PM—after which the lighting system may be switched off by the timed switching system.

As an alternative to photosensors, consider installing a microprocessor-based timed switching system for controlling outdoor lighting. Systems are available that predict seasonal dusk and dawn switching times and automatically switch the outdoor lighting systems according to this schedule. Such systems have a back-up battery and program memory to ensure that the "solar schedule" will remain properly programmed in the event of a power failure. Microprocessor-based daylight switching systems can incorporate "pre-dawn" scheduled switching functions. Many systems provide the capability to program various lighting schedules over a multi-year period.

Buyer Beware

Mechanical time clocks are not recommended for daylight switching control because they can be relatively inaccurate in scheduling on/

Figure 11-15. Photocells are used for automatically switching outdoor lighting systems on at dusk and off at dawn. *Courtesy: Paragon Electric Company, Inc.*

off functions, and may get "off schedule" if not properly maintained. Photocells should be properly calibrated and maintained to eliminate wasteful "day-burning."

Daylight switching *indoors* has been applied with varying degrees of success. In relatively low mounting heights, users may object to the use of automatic switching of the lighting system during daylight hours because it draws attention to sudden changes in illumination. However, adverse occupant reactions can be minimized if the sensor can be programmed to turn on the lights when the ambient light level drops to about 30 fc and turn off the lights when the ambient light level climbs to about 65 fc. This strategy will minimize rapid cycling of the lights. Still, the most successful indoor applications for daylight control usually involve *dimming* instead of switching. Some occupancy sensors provide daylight switching control in conjunction with their occupancy switching control. A trial installation is recommended to assess user acceptance of this technology.

One application where indoor daylight switching may prove both cost-effective and pleasing to occupants involves the installation of an *active daylighting system*. Using motorized sun-tracking mirrors, sunlight is reflected through skylights and distributed throughout the building interior via a wide-angle diffuser. An automatic control switches off the artificial lighting when sufficient daylighting is provided by the system. Available as new and retrofit systems in existing single story structures, these systems can provide sufficient illumination during sunny days (and bright cloudy days), thereby eliminating the need for artificial lighting during most daylight hours.

Chapter 12

Dimming Controls

D imming controls provide facility managers with an effective means to achieve energy cost savings and flexibility: Users can select their own light levels, and light output can be automatically reduced based on ambient conditions such as availability of daylight. In this chapter, we will review dimming control options and the strengths and weaknesses of various upgrade strategies involving dimming.

DIMMING CONTROLS

Dimming controls can achieve much more than reductions in energy costs. Their flexibility can also improve lighting quality and worker productivity by matching light output to changing visual task requirements, user-specific visual capabilities and daylight levels. Because traditional static systems are designed to deliver the appropriate lighting for all occupants under the *worst* anticipated conditions (no daylight, uncleaned fixtures and older occupants working on critical tasks), such systems waste energy by providing more lumens than needed, most of the time. By matching lumen output to meet the lighting requirements at any given point in time, dimming controls will reduce power requirements, which can yield significant cost savings for both energy use *and* electrical demand. The introduction of advanced electronic dimming products has reduced the cost and complexity of installing a retrofit dimming control system.

This chapter addresses several types of dimming systems. The most familiar type is continuous, where the light level varies gradually as the requirements change. *Step* dimming is another type of dimming where the light output jumps from one level to another as the requirements change. Most step-dimming technologies provide between two and five levels of light output. The final dimming category is *static* dimming, which refers to technologies such as power reducers that provide a fixed reduction in light output.

OVERVIEW OF DIMMING STRATEGIES

There are many reasons to consider dimmable lighting systems. Although most of these strategies are intended for energy savings, some will provide flexibility for adapting the lighting system to meet individual preferences. Following this overview, these dimming strategies are discussed in more detail.

Daylight Dimming
Depending on the amount of window area in working spaces, daylight may be sufficient to supply some or all of the lighting needed for performing visual tasks. However, daylighting alone does not save energy unless the electric lighting system is controlled. Daylight-dimming strategies involve the use of a photosensor that provides a signal to the lighting system, which automatically reduces electric light output during times when daylight is available. In addition to their potential to save energy costs, daylight-dimming controls can also save on electrical demand charges, as peak daylight levels often coincide with peak electrical demand periods. Daylight-dimming systems also provide lumen maintenance control as described below.

Lumen Maintenance Control
The goal of lumen maintenance control is to provide a *constant* light level, compensating for changing lumen depreciation effects. The same light sensors used in daylight dimming are used in lumen maintenance control. The sensors detect the gradual reduction in light output due to accumulated dirt and aging lamps, and the control system responds by gradually increasing input power to maintain the desired light level. Lumen maintenance controls save energy by limiting the input power when the lamps are new and the fixtures are clean. After installing new lamps and cleaning the luminaires, the manual adjustment on the photosensor should be tuned to lower the illuminance by at least 25-30 percent—the amount of lamp lumen depreciation and luminaire dirt depreciation to be expected during the maintenance cycle. In order for a lumen maintenance control strategy to save energy, the luminaires must be cleaned and relamped on a regular basis.

Manual Dimming

Using manual dimming controls, the light output from individual luminaires or groups of luminaires can be reduced to match the area's visual requirements. Where daylight-dimming and/or lumen maintenance controls have been installed, the maintained light level can be adjusted using the manual control built into the photosensor. Other methods for manually adjusting light levels involve the use of wall-dimmers, hand-held remote controllers, and LAN-based user controls.

Scheduled Dimming

In many applications, a lighting control system is programmed to change the light levels according to the time of day. For example, in a retail application, the lights may be operated at lower levels during routine stocking and maintenance operations when the store is closed, and at higher levels when the store is open for customers.

Occupancy-sensed Dimming

In some areas, it may be more appropriate to dim (rather than turn off) the lights where the occupancy sensors indicate vacancy. One application of occupancy-sensed dimming is in warehouse aisles with HID lighting; instead of switching the lamps off, a bi-level capacitive switching system avoids HID restrike and warm-up delays by dimming the HID luminaires down to a standby level during periods when aisles are unoccupied. When people enter the aisles, the HID systems take very little time to return to full light output.

DAYLIGHT DIMMING

Daylight dimming is the most common fluorescent dimming strategy used for energy management. The simplest fluorescent daylight dimming systems consist of photosensors that are wired directly to controllable (dimmable) electronic ballasts. Some manufacturers provide a photosensor to control every ballast, while others provide a single photosensor that can control many ballasts simultaneously. Still others require that an integrated controller be wired between the sensor and the ballasts. Figure 12-1 illustrates these three unique configurations.

Because the control wiring between the photosensors and the bal-

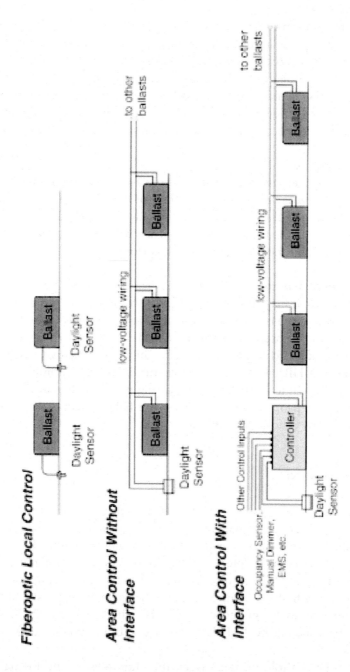

Figure 12-1. Typical configurations of fluorescent daylight-dimming systems. The more complex configurations tend to be more expensive. *Courtesy: EPA Green Lights.*

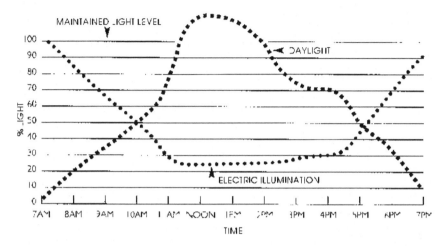

Figure 12-2. Daylight-dimming systems can maintain a constant light level on visual tasks by adjusting fluorescent light output to compensate for the changing amounts of ambient light. *Courtesy: National Lighting Bureau.*

lasts can be located in the plenum above a suspended ceiling, there is no requirement to rewire the existing power circuits. The daylighting "zone" (consisting of the luminaires to be dimmed) is defined by these low-voltage connections which are *independent* of the power circuits. Therefore, even if power circuits are oriented perpendicular to the windows, the daylight dimming system can be installed to dim only those fixtures next to the windows.

"Closed-loop" photosensors are used in daylight dimming systems for maintaining a constant light level. These sensors measure the contributions of daylight and artificial lighting and continuously adjust the output of the electric lights so that the illumination is maintained at the set-point level.

A manual adjustment on the photosensor allows users to select the light level to be maintained in both the absence of daylight and during the dimming process. Most controllable electronic ballasts used for energy management will reduce light output down to about 20 percent, with others down to 10 percent. When this minimum output level is reached, increasing daylight contributions may further elevate light levels beyond the manually adjusted set-point.

Application Guidelines

Ceiling-mounted photosensors should be installed at a specific distance from window areas, generally two-thirds of the width of the zone where the daylight dimming luminaires are located. Because the photosensor is sensing the brightness of the surfaces within its angle of acceptance (viewing range), it should be mounted above a representative work area. Ask for manufacturer data about the acceptance angles of the photosensors under consideration.

Although daylight-dimming can be cost-effective in many retrofit applications, it is most cost-effective in new construction or major remodeling projects. In such applications, dimming can be planned and integrated with other factors such as building orientation, glazing and energy management systems.

Buyer Beware

Proper placement of photosensors is critical to the success of a daylight dimming installation. Follow manufacturer specifications carefully.

If architectural structures or partitions reduce the amount of available daylight in selected spaces within the daylighting zone, exclude the affected luminaires from daylighting control. Alternatively, if daylighting contributions vary widely within the daylighting zone, consider installing a local-area daylighting system that provides a photosensor for each luminaire.

The use of blinds can dramatically reduce (or eliminate) the savings expected from daylight dimming. Window film should be used in place of blinds wherever feasible. Where blinds are used, however, window glare can be controlled while allowing daylight to enter the space: Simply orient the blades so that daylight is reflected to the ceiling.

To achieve sustained energy savings, be sure to adjust the photosensor so that the proper light levels are maintained. To calibrate the photosensor, follow these steps:

1. When daylight levels are low or nonexistent, reduce the light output by adjusting the tuning control on the photosensor to a point that is *below* the target workplane illuminance. (Have a light meter placed on the work surface.)

2. Use the photosensors' tuning control to *increase* the light output until the illuminance reaches the desired light level.

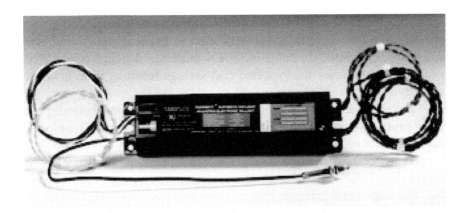

Figure 12-3. Retrofit daylight-dimming installations can be affordable when tandem-wiring 4-lamp dimming electronic ballasts and using integrated light sensors. These systems can be manually adjusted to provide location-specific light levels (tuning). *Courtesy: Flexiwatt, LCC.*

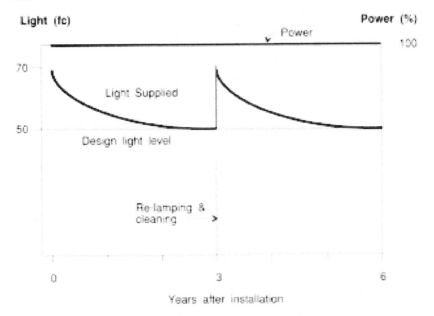

Figure 12-4. In conventional lighting systems, input power remains constant as light output drops over time. The result is overlighting and wasted energy during most of the maintenance cycle. *Courtesy: California Energy Commission.*

When calculating energy cost savings expected from a dimming system, take into account the specific electric demand charge and rate structure; some rate schedules include a ratcheted demand charge that could reduce or delay the expected cost savings resulting from lowered peak demand.

Note that the photosensor, controller (if needed) and ballast must all be mutually compatible for the dimming system to work. These components must be specified as a system, not individually.

When cost-justifying a daylight dimming system, don't neglect to include the labor costs of engineering design and system commissioning. The engineering design step is more complex when laying out an area-control system, including positions of sensors, wiring plans for controllers and low-voltage circuit layouts for luminaire control. Commissioning includes the process of tuning the sensors to provide the desired maintained illumination. Maintenance staff must be trained to properly troubleshoot and repair any dimming-related problem.

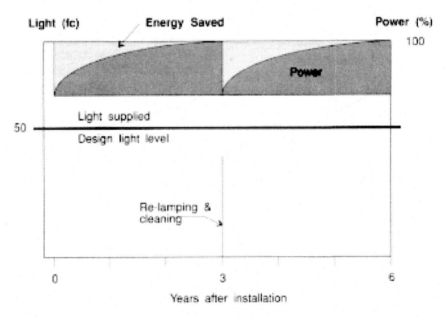

Figure 12-5. Lumen maintenance controls improve lighting quality by maintaining constant light levels, compensating for lumen depreciation effects with gradually increased power input. Energy is saved by eliminating overlighting. *Courtesy: California Energy Commission.*

To *directly* measure the savings achieved with a dimming system, refer to Chapter 18.

LUMEN MAINTENANCE CONTROLS

With lumen maintenance controls, light levels will remain constant—instead of declining—between luminaire washing and relamping cycles. Properly tuned, these controls will maintain the *correct* light levels, rather than allowing the system's output to gradually decline from excessive light levels to potentially inadequate light levels. The energy savings result from eliminating the overlighting that is often necessary to compensate for the lumen depreciation that occurs in all lighting systems.

Application Guidelines

Because lumen maintenance controls operate on the same equipment and on the same principles as daylight dimming, the guidelines for using lumen maintenance controls are virtually the same as daylight dimming. The primary difference is that light sensors are not compensating for daylight contributions; the system compensates only for light output reductions due to luminaire dirt depreciation and lamp lumen depreciation. Ambient light sensors should be positioned over visual task areas to determine the approximate light levels that are delivered by the luminaires.

With lumen maintenance controls, energy savings are *only* achieved by cleaning and relamping the luminaire. Users who install these controls will have a financial motivation to implement a program of regular luminaire cleaning and scheduled relamping— under lumen maintenance control, the cleaner the luminaires are (and the newer the lamps are), the less energy the system will use.

Additional savings can be attained with lumen maintenance controls by manually adjusting the light output of the system to meet individual needs. In many cases, occupants will desire lower light levels than are provided by the base system.

Buyer Beware

In very clean environmental conditions (such as in modern office buildings), dirt build-up in fluorescent troffers may not occur at a rapid enough pace to cost-justify lumen maintenance controls. Furthermore, new triphosphor fluorescent lamps—particularly those with a color rendering index of 82-85—lose only 4-7 percent of their initial rated light output over their lives. The most cost-effective applications of this strategy are in new construction, in clean to moderately dirty environments, or where tuning control is desired (see below).

MANUAL DIMMING (TUNING)

When considering lighting upgrades for a typical commercial building, the facility manager has a difficult decision to make: What light level should be maintained? Usually a conservatively high footcandle is specified, because it is less risky to have too much light than not enough. Still, enormous amounts of energy are wasted with this approach. Others will choose to specify a number of different kinds of electronic ballasts, each with different ballast factors so that unique space types will be illuminated to the task-specific light level—until the defined task changes.

Manual dimming strategies efficiently deal with these uncertainties while ensuring that all occupants receive the light levels that they require to most effectively perform their work. In addition, when workers are given some control over their specific workstation illumination, they are more likely to accept the new lighting system, appreciating its ability to meet their individual preferences for light levels.

Application Guidelines

In retrofit applications, the most common type of manual dimming control comes in the form of electronic dimming ballasts that can be dimmed from wall control devices without the need for any additional control wiring. These products are normally applicable to private offices and conference rooms.

As an alternative to wallbox dimming, wireless control technologies are also available from selected ballast manufacturers that can be used in retrofit projects. A sensor is installed in the ceiling that is connected directly to the ballasts using low-voltage control wiring. The sensor receives wireless dimming signals from wireless transmitters and forwards the command to the connected ballasts. In addition to providing manual dimming control, some sensors can also sense occupancy and daylight conditions within the space, thereby offering a complete, localized control solution.

In renovation projects where new fixtures will be selected, more advanced forms of personal dimming control become cost-effective. For example, when selecting luminaires for lighting open-office worksta-tions, consider the "workspace-specific" application of individual sus-pended direct-indirect luminaires: Using wireless transmitters or PC-addressable dimming technology, the downlighting component from these luminaires can be dimmed by individual users to suit personal preferences and needs.

Buyer Beware

Trial installations are recommended before proceeding with manual dimming upgrades. Occupants and/or maintenance staff should be trained on the proper use of the controls. In open office applications, a trial installation would demonstrate if individual luminaire tuning would cause an aesthetic problem with varying brightness among the luminaires.

HID POWER REDUCERS

HID power reducers provide permanent reductions in light output with reductions in power consumption. HID power reducers are de-signed to be wired in conjunction with the HID ballast *or* they can be installed to control an entire circuit.

Application Guidelines

Power reducers are designed to achieve a preset light output reduc-tion—and energy savings—of 20-25 percent. In addition, power reducers extend HID ballast life by reducing ballast operating temperature. They should be considered as economic alternatives to panel-level HID dim-ming systems if variable control of light output is not needed.

Buyer Beware

Power reducers are typically designed to work only with the more common CWA ballasts and HID lamp wattages of at least 175 watts. Lamp types that can be controlled include mercury vapor, metal halide and HPS. If permanent reductions in light output are not required, con-sider installing reduced-wattage (reduced-output) HID "energy-saver" lamps; these lamps are a low-cost retrofit, but are subject to "snap-back."

A New Paradigm for Dimming Control: DALI

Common to all *conventional* dimming controls is the broadcasting of dimming commands over a hardwired control circuit. Since the advent of the Digital Addressable Control Interface (DALI), the rules for dimming control have changed. Using DALI controls, specific commands can be addressed to individual ballasts, which improve the flexibility of dimming while minimizing the number of circuits required.

DALI is an international standard that establishes a common language for communicating between digital control devices (wall controls, user PCs, etc.) and DALI-compliant ballasts and relays. This non-proprietary protocol opens the door to multiple manufacturers producing dimming system components that are interchangeable and are simply connected to each other on a 2-wire DALI "loop" or "communications bus".

DALI-compliant ballasts store control settings in memory and respond to digital commands addressed to them. These settings include a unique address, the current light output value, maximum and minimum light output values, group number(s), scene-specific light output values, fade rate, fade time and many others. The settings for DALI addressable relays are similar to DALI ballasts, except that the light output values are confined to "level 0" (off) and "level 254" (on). DALI settings can be modified using the following types of control equipment:

Group and Scene Controllers: DALI group controllers are wallbox devices that can dim or switch a specified group of ballasts assigned to a DALI group. DALI scene controllers are wallbox devices that send out a "go-to-scene #" command, and all devices that have that scene number (0-15) and corresponding light output value stored in memory will fade to that level. In addition, conventional ceiling-mounted occupancy and daylight sensors can be connected to DALI group controllers to automatically modify light output values based on room conditions.

Network Control Systems: Using the building's LAN system, individual users and system administrators can send messages to a server/database, which routes the command to the intended address(es), subject to permissions established in the database. In addition to providing individual user PC control by assigned addresses, the administrator can implement building-wide control strategies including load shedding and time-based scene switching. Finally, DALI's two-way communication feature allows the administrator to poll any selection of DALI ballasts in the building to identify lamp failures and to monitor energy consumption.

Check with the lamp and ballast manufacturers to determine if the installation of power reducers will have any effect on their warranties.

Although some of the product literature claims little or no perceived reductions in light output, these devices *will* reduce light output. Trial installations are suggested to measure light output reductions and energy savings. Ask the manufacturer to arrange a 60-day trial installation for evaluating the performance in a specific application. (Insist on a risk-free installation where if the product does not perform as claimed, it will be removed at no cost.) To determine the performance of the system, measure the average light level (in footcandles) and electrical demand (in kilowatts) of the lighting circuit prior to installation, and make the same measurements after the installation is in place. Refer to Chapter 18 for guidance in performing and evaluating trial installations.

PANEL-LEVEL HID DIMMING

Panel-level HID dimming is a strategy for uniformly controlling all HID (or in some cases fluorescent) luminaires on designated circuits. A control system is installed at the electric panel that reduces the power supplied to the circuit.

Methods
Continuous dimming is accomplished using one of these methods:

Variable-voltage Transformer
Reducing voltage to the circuit causes reductions in light output and energy input. Typically used with existing CWA ballasts, these systems enable reductions down to about 50 percent of rated power.

Variable Reactor
By reducing current to the circuit while maintaining constant voltage, a wider range of dimming is possible—down to about 30 percent of rated power.

Waveform Modification
Using electronic components, the incoming AC voltage and current waveforms are altered, allowing dimming down to about 50 percent of rated power.

Application Guidelines

Circuit dimming can be controlled manually or by inputs from occupancy sensors, photosensors, time clocks or energy management systems. For example, using photosensors in a warehouse with skylights, the HPS lighting system could be uniformly dimmed in response to the available daylight, saving substantial amounts of energy.

Other applications involve wholesale merchandising outlets that require higher light levels during normal business hours, and reduced light levels during routine maintenance and stocking operations. The scheduling control system can automatically adjust the light levels based on the business operating schedule.

Buyer Beware

Slight reductions in efficacy result from the dimming of HID systems. Light output reductions are about 1.2 to 1.5 times the power reduction in metal halide systems and about 1.1 to 1.4 times the power reduction in HPS systems. Manufacturers can provide the specific lumen-wattage performance curves for the specific systems being controlled.

Figure 12-6. Variable-voltage transformers can be used to uniformly dim HID lighting systems and save energy. *Courtesy: Superior Electric.*

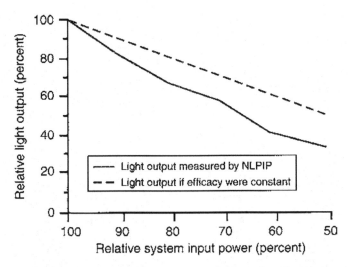

Figure 12-7. Note that as HID systems are dimmed, efficacy is reduced. This curve indicates the relationship between light output and power input for a 400W metal halide system as it is dimmed. *Courtesy: Lighting Research Center's National Lighting Product Information Program. ©Rensselaer Polytechnic Institute.*

Note that some panel-level dimming systems are incompatible with electronic ballasts. Check with the manufacturer to determine if their dimming system is compatible with electronic ballasts and whether the system introduces harmonic currents.

Dimming HID lamps below 50 percent power may result in a significant reduction in lamp life and may void lamp warranties. In addition, dimming HID lamps can make flicker more visible, especially in HPS lamps. Unlike fluorescent lamps, metal halide lamps tend to shift to a higher color temperature (cooler appearance) as they are dimmed. This effect can be minimized with the use of coated metal halide lamps.

CAPACITIVE-SWITCHING HID SYSTEMS

Capacitive-switching HID systems are designed to provide either bi-level or tri-level HID illumination based on inputs from occupancy sensors, manual switches, photosensors or scheduling systems. When reduced light output is desired, the system switches to a second (or third) capacitor that reduces the system's electrical current and saves energy. Ca-

Figure 12-8. Bi-level HID switching systems can save energy by providing a low level of "standby" light output while the space is unoccupied. *Courtesy: Thomas Lighting, Inc.*

pacitive-switched dimming can be installed as an energy-saving retrofit to existing HID luminaires or as a direct luminaire replacement.

Capacitive-switching HID upgrades can be less expensive than installing panel-level HID dimming systems, especially in circuits with relatively few luminaires. In addition, it allows for control of individual luminaires, rather than entire circuits.

Application Guidelines

The most common applications of capacitive switching are occupancy-sensed bi-level control in parking lots and warehouse aisles where there are long periods when the space needs to be lighted but is unoccupied. Due to the relatively long restrike and warm-up times associated with HID systems, it may not be feasible to switch HID lighting off and on with the use of occupancy sensors. With capacitive switching systems, the occupancy sensor will detect motion and send a signal (typically by powerline carrier, low-voltage wire or fiber-optic cable) to the bi-level HID control system. The system will respond by rapidly bringing the light levels from a standby reduced level to about 80 percent of full output, followed by a short warm-up time between 80 percent and 100 percent of full light output. Alternatively, some new luminaires are sold with a dedicated occupancy sensor and capacitive-switching ballast so that no additional control wiring is required.

Depending on the lamp type and wattage, the standby lumens are roughly 15-40 percent of full output and the standby wattage is 30-60 percent of full wattage. Therefore, during periods that the space is unoccupied and the system is dimmed, energy savings of 40-70 percent are achieved.

Tri-level capacitive switching involves switching between three capacitors to provide three levels of illumination. This degree of illumination flexibility may be warranted in multipurpose rooms where the use of a continuous HID dimming system may not be cost effective.

Buyer Beware

Lamp manufacturers do not recommend dimming below 50 percent of the rated input power. Check with lamp suppliers to determine whether the bi-level system will affect their lamp warranties. As opposed to reducing voltage or modifying waveforms, capacitive dimming (reducing current) allows the greatest amount of dimming with minimal effects on lamp life.

ELECTRONIC DIMMING HID BALLASTS

Electronic dimming HID ballasts can now be found in new, dimmable HID luminaires. In addition to improved efficacy, these ballasts are lighter, provide better color control (at least at full output) and produce less stroboscopic effect. To preserve lamp life, the dimming range with electronic ballasts is typically limited to 100-50 percent light output with metal halide and 100-30 percent light output with HPS. Harmonic distortion is usually well under 20 percent.

Application Guidelines

Because these ballasts are typically not cost-justified for retrofit applications, they are available in new luminaires. Like controllable fluorescent ballasts, these units will accept inputs from compatible light sensors, occupancy sensors, load scheduling systems and manual controls. Using low-voltage wires, control circuits can be independent of power circuits.

Buyer Beware

Because this is a new technology, buyers should insist on evaluating this technology in a trial installation. Warranty support should be an important criterion in purchase decisions. The same limitations that apply to controllable fluorescent ballasts also apply to these new HID controllable ballasts.

Chapter 13

Upgrades to Modify Light Levels

O ne of the most common mistakes made in lighting upgrade design is simply assuming that the existing light levels should be maintained. Before selecting upgrade technologies for a lighting application, first determine the light level needed to perform the visual tasks in the space. This first application chapter presents the technology options for adjusting lighting system output to meet target illumination levels.

MODIFYING LIGHT LEVELS

Overlighting wastes energy and contributes to visual discomfort, while underlighting can cause productivity and morale to suffer. Therefore, it is critical to establish the appropriate target light level and design the lighting upgrades accordingly.

Although some organizations have developed their own illuminance standards, the lighting industry generally follows the recommendations published by the IESNA. Refer to Chapter 1 for guidance in selecting the appropriate maintained illuminance value.

Existing light levels can be measured with an illuminance meter. However, note that the light levels produced by the system are constantly changing due to aging lamps and dirt accumulation. Therefore, it can be difficult to know precisely what the average maintained footcandles are in a particular application. Refer to Chapter 18 for guidelines in measuring footcandles under controlled conditions and for using software to calculate footcandles.

Before modifying light levels, consider the light loss factors that affect the existing system output, including lamp lumen depreciation (aging lamps) and luminaire dirt depreciation (accumulated dirt on optical surfaces). The IESNA recommendations are based on *maintained* levels—the average footcandles expected between lamp replacement and fixture cleaning operations. Refer to Chapter 20 for a complete discussion of these light loss factors.

Reducing Light Levels In Overlighted Spaces

Reducing light levels in overlighted spaces will not only save energy, but can also improve lighting quality. *More lighting is not necessarily better lighting!* Excessive light levels can contribute to glare, headaches and fatigue. In general, light levels should be re-assessed as the IESNA-recommended light levels are revised to generally lower values.

Following are examples of commonly overlighted space types:

Offices With Computers

In past decades, office lighting generally consisted of 2 ft. x 4 ft. 4-lamp troffers that provided uniform general lighting of over 100 fc. Higher light levels were warranted as office workers struggled with relatively low-contrast pencil-based visual tasks. However, with today's higher-contrast printed documents, modern office lighting systems should only provide 30-50 fc for high-quality illumination.

Circulation Area

The visual tasks in hallways and corridors consist of general navigation or "collision avoidance." The IESNA-recommended illumination for these tasks would be 5-10 fc, depending on whether visual tasks are occasionally performed in these areas. In many cases, the existing circulation lighting can be reduced by 50 percent or more and still exceed IESNA illuminance recommendations.

Automation

The use of automated manufacturing techniques has replaced the need for high-intensity lighting previously used for human-based assembly tasks. Now, these spaces require a lower level of lighting for monitoring equipment operation. In some applications, *all* tasks are performed robotically and *no* illumination is needed.

In applications where the visual task or the illumination requirements have changed since the lighting system was originally designed, consider the following methods for reducing light levels. *For more information about these upgrades, refer to Chapters 4, 6, 8 and 12.*

Table 13-1 provides an example of how fluorescent technology options can be applied to reduce luminaire light output and save energy. This example features just a few of the technologies described in this chapter for reducing light levels.

Delamping

Delamping is a very simple method for reducing light levels in spaces with multilamp luminaires. In general, delamping reduces light levels in proportion to the number of lamps removed. To ensure permanent savings, disconnect and remove the unused ballasts and lamp sockets. Delamping may be combined with the use of higher output lamps and/or ballasts, reflectors, lens upgrades, luminaire cleaning and/or task lighting to partially offset reductions in light levels and improve system efficiency.

Table 13-1. Selected 2x4 system choices for reducing light levels.

Lamp-Ballast System Description	Source Lumens[1]	Luminaire Efficiency[2]	Relative Wattage	Relative Light Output[3]
4-lamp T12 EE-magnetic (0.94 BF)	9,977	68%	100% (base)	100% (base)
4-lamp T12/ES EE-magnetic (0.94 BF)	8,023	68%	82%	80%
4-lamp T8 electronic (0.75 BF)	7,781	70%	57%	80%
2-lamp T8 electronic (1.28 BF)	6,639	74%	49%	72%
2-lamp T8 electronic (0.88 BF)	4,565	74%	35%	50%
2-lamp T8 electronic (0.75 BF)	3,890	74%	31%	42%

[1]Cool white T12 lamps and 75 CRI T8 lamps; includes effects of lamp lumen depreciation.
[2]Lensed 2x4 luminaire.
[3]Includes effect of luminaire efficiency.
Notes: EE = Energy-Efficient, ES = Energy Saver, BF = Ballast Factor.

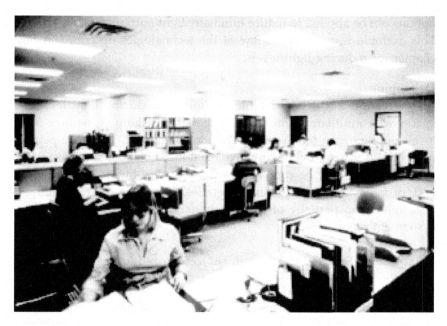

Figure 13-1. Before computer use became widespread, office lighting systems provided over 100 fc with minimal glare shielding. *Courtesy: IESNA, New York City.*

When *parallel*-wired electronic ballasts are used in 3-lamp or 4-lamp luminaires, selective delamping can reduce light output to meet occupant preferences or reduced task illumination requirements in specific locations. For example, a single lamp can be removed from a 4-lamp parallel-wired ballast circuit, and the light output will drop by close to 25 percent as the other lamps continue to illuminate (at a slightly higher output). Check with the electronic ballast manufacturer to determine if selective delamping will have any adverse effects on lamp or ballast life.

Reduced-Output Electronic Ballasts

Reduced-output electronic ballasts operate fluorescent lamps at the same high efficacy as other electronic ballasts, but with specified reductions in both light output and energy consumption. As an alternative to delamping, these ballasts keep all lamps operating at reduced output, resulting in improved luminaire appearances. Electronic ballasts with a reduced ballast factor (0.67-0.80) reduce light output and increase energy savings—without costing more than full-output electronic ballasts.

Reduced-output Lamps

Energy-saver T12 and T8 fluorescent, 17W 2 ft. T8 lamps, energy-saver HID lamps and reduced-wattage compact lamps can be used to reduce light levels and save energy.

Energy-Saver T12 Fluorescent Lamps

Although the 4 ft. "energy-saver" 34W fluorescent lamps reduce fluorescent light output by up to 20 percent, they do not affect system efficacy. Instead, consider using reduced-output electronic ballasts (as described above) with 32W T8 lamps because they reduce system light output *and* improve system efficacy. Note, however, that the system efficacy *is* improved when using the *8 ft.* "energy-saver" 60W T12 lamps— particularly when triphosphor lamps are used with electronic ballasts.

Energy-Saver T8 Fluorescent Lamps

What can facility managers do when they have already invested in full-output electronic ballasts and T8 lamps, and light levels are too high? Similar to energy-saver T12 fluorescent lamps, lamp manufacturers offer a reduced-wattage alternative for T8 systems that offers reduced lumen output. Energy-saver T8 fluorescent lamps come in wattages as low as 28 watts and deliver a reduction in light output that corresponds to the reduction in power. Note, however, that these lamps require a higher minimum starting temperature (60 degrees), and they are not suitable for dimming or use with ballasts with a very low ballast factor (under 0.7).

17W T8 Lamps

One option for reducing light output from 2x2 fluorescent luminaires is to replace the 34W or 40W T12 U-lamps with straight, 24-inch, 17W T8 lamps. Using a UL-listed assembly, the existing sockets, raceways and ballast are replaced with a 2-lamp or 3-lamp F17T8 electronic system. Lumen output reductions can be partially offset with the use of retrofit reflectors. These upgrades yield improvements in system efficacy and luminaire efficiency.

Energy-Saver HID Lamps

Major manufacturers of high-intensity discharge lamps have introduced several models of reduced-wattage metal halide and HPS lamps. These lamps are direct replacements and produce up to 18 percent reductions in light levels and energy consumption.

Reduced-wattage Compact Lamps

In compact luminaire applications, there are many wattages of compact fluorescent and halogen lamps from which to choose. Select the appropriate lumen output package that will deliver the required illumination without overlighting the visual task. When evaluating compact fluorescent light levels, however, remember that their light output will depreciate by up to 20 percent over their life.

Manual Dimming Controls

With wide variations in worker age and visual task requirements, the flexibility offered by manual dimming controls can yield impressive energy savings while optimizing light levels for individual workers. Fixed-output systems are typically designed to deliver the light level required for the worst-case situation—lighting the most demanding visual task anticipated, to be performed by the occupant with the lowest visual capabilities.

Manual Dimming of Fluorescent Systems

The sales of dimmable electronic fluorescent ballasts are increasing as new control technologies are introduced. In retrofit applications, manual dimming controls can be installed that do not require low-voltage control wiring, such as line-voltage dimming electronic ballasts and multi-level electronic ballasts designed for dual-switching circuits.

Other methods for fluorescent dimming control include 0-10 Volt *analog* control and DALI (Digital Addressable Lighting Interface) *digital* control. Both systems offer options for wallbox dimming control. However, DALI also offers the option for PC-controlled dimming. In addition, wireless solutions are available that enhance the convenience of personal dimming.

Panel-level Dimming Systems

Panel-level dimming systems can be used for dimming high-intensity discharge systems and some fluorescent systems. These systems provide uniform dimming of all luminaires on designated circuits. Dimming can be controlled manually or by inputs from occupancy sensors, photosensors, time-clocks or energy management systems.

Step-dimming HID Systems

As an alternative to continuous-dimming with panel-level systems, step-dimming of HID systems can be achieved with capacitive switching ballasts. Using multiple capacitors, bi-level and tri-level dimming is achieved by reducing current to the HID lamps.

Power Reducers

Power reducers are devices that provide fixed reductions in light output and power consumption of magnetically ballasted HID or fluorescent systems. They can be installed as ballast add-on devices, or they can be installed at the electrical panel to control one or more circuits.

How *Not* to Reduce Light Levels

One very short-sighted approach to dealing with the discomfort of excessive illumination is to replace the shielding material with less-efficient products (such as tinted lenses or small-cell parabolic louvers). These solutions do not take advantage of the energy cost savings that can be easily achieved by the alternatives described above.

INCREASING LIGHT OUTPUT
IN UNDERLIGHTED SPACES

Inadequate light levels can hamper productivity and safety. In these cases, install the most efficient systems for boosting light output to the target levels. *The gain in efficiency can usually produce significant increases in light levels while yielding energy cost savings.*

Table 13-2 illustrates how technologies can be chosen to improve light levels. These technologies and other approaches are described below. *For more detailed discussions of these upgrades, refer to Chapters 4, 6 and 8.*

Higher Output Ballasts

In addition to partial-output and full-output electronic ballasts, *extended output* electronic ballasts are also available. Typically, these higher-wattage ballasts have a ballast factor in the range of 1.0 to 1.3 for boosting light output while maintaining the high efficacy that is characteristic of electronic ballasts.

Higher Output Lamps

Higher output lamps include high performance T8 lamps and position-specific metal halide lamps.

High Performance T8 Lamps

Lamp manufacturers offer higher-grade versions of their T8 lamps that offer improved lumen output, longer life, improved color rendering index (85+) and improved stability in dimming and low-temperature applications. Compared to the standard genre of T8 lamps, these lamps offer up to 10 percent higher lumen output, with rated lamp life extended from 20,000 hours to 24,000 hours.

Position-Specific Metal Halide Lamps

Higher output metal halide lamps are available that are designed to be operated in a specific orientation. If "universal" lamps are currently in use, these position-specific lamps may improve lumen output by up to 25 percent.

Task Lights

Instead of increasing the light output from the ambient (overhead) lighting system, consider installing task lights to increase task illumination. Because occupants have varying needs for illumination based on age and visual task requirements, it is usually more cost-effective and energy-efficient to provide compact fluorescent task lights as needed at individual workstations. If certain workers require higher task light levels, higher-output (or additional) task lights may be furnished. In industrial applications, task lighting can take the form of increasing the number of low-bay luminaires over a specific work area.

Increase Luminaire Efficiency

One of the primary causes of inadequate illumination is due to aging or poorly maintained lighting systems. By increasing luminaire efficiency (the percentage of bare-lamp lumens that exit the fixture), resultant light levels will increase. Below are some steps to improving luminaire efficiency.

Replace Inefficient Lenses and Louvers

Lenses that are either translucent-white or are acrylic and have yellowed can trap a large percentage of lumens within the fixture. In addition, small-cell louvers (1/2-inch to 2-inch cells) can make sacrifices in efficiency to provide glare shielding. By replacing these components with new acrylic lenses or deep-cell parabolic louvers (cells greater than 4 inches across), light levels can be enhanced.

Use Open HID Luminaires

In metal halide applications, luminaire efficiency may be improved by using lamps that are rated for open luminaire operation. When these lamps are used, the enclosure/lens can be removed (which can absorb some of the light), thereby improving the luminaire's light output.

Retrofit with Specular Reflectors or White Powder-coat Reflectors

Retrofit reflectors may be the most cost-effective option for restoring the performance of reflector surfaces in older luminaires where the finish may be dull or deteriorated. By retrofitting with high-reflectivity aluminum, silver-film or white powder-coat reflectors, light output from older luminaires can be increased by well over 20 percent. For HID luminaires, retrofit specular and clear reflectors are available. Trial installations are recommended to determine reflector performance.

Periodically Clean Luminaires

Accumulated dirt on lamp and luminaire surfaces can reduce light output by over 40 percent. By cleaning luminaires whenever lamps are re-

Table 13-2. Selected 2×4 system choices for increasing light levels.

Lamp-Ballast System Description	Source Lumens[1]	Luminaire Efficiency[2]	Relative Wattage	Relative Light Output[3]
2-lamp T12/ES/CW EE-magnetic (0.94 BF)	4,012	72%	100% (base)	100% (base)
2-lamp T12/CW EE-magnetic (0.94 BF)	4,989	72%	122%	124%
2-lamp T10 EE-magnetic (0.92 BF)	6,059	73%	129%	153%
2-lamp T8 electronic (1.28 BF)	7,261	74%	119%	186%
2-lamp T8 electronic (1.28 BF) with reflector	7,261	84%	119%	211%

[1]Cool white T12 lamps, 82 CRI T10 lamps and 85 CRI T8 lamps; includes effects of lamp lumen depreciation.
[2]Lensed 2×4 luminaire.
[3]Includes effect of luminaire efficiency.
Notes: EE = Energy-Efficient, ES = Energy Saver, CW = Cool White, BF = Ballast Factor.

placed, average light levels can be increased. However, in dirty environments, it may be necessary to clean the luminaires more frequently.

Retrofit or Replace?

Using today's retrofit lighting technologies, it is possible to upgrade lighting systems to modify light levels and improve energy efficiency without having to replace the luminaire. However, many users may find that it can be more cost-effective to purchase and install new, energy-efficient luminaires than to "rebuild" existing luminaires. To restore the performance of relatively old luminaires, the upgrade may involve removing the lamps and ballasts, disconnecting and/or relocating the lamp sockets, and installing a combination of new ballasts, reflectors, lamps and lenses or louvers. Upgrading with new luminaires—that yield the same efficiency and visual comfort—may be cost-justified on installation labor savings alone.

When to retrofit? In most lighting upgrades, retrofitting the existing luminaire will be the most cost-effective method for maximizing energy savings while maintaining or improving lighting quality. This is particularly true in applications where the luminaire optical surfaces (reflector and lens/louver) are in relatively good condition, and simple lamp/ballast replacements are the only cost-effective upgrades. Even when low-glare lens upgrades and/or reflectors are added, the overall cost could be less than purchasing and installing new luminaires.

In addition, when a building is identified as having asbestos in the ceiling structure, it is usually cost-prohibitive to replace the luminaire, which would expose the asbestos. In these situations, consider installing a retrofit package that upgrades the lamps, ballasts and/or reflector (and lens, if needed) without removing the luminaire.

When to replace? The installation of new luminaires is generally favored in situations where the luminaire lens and reflector are deteriorated or where a new system is desired for renovation or improved visual comfort. For example, if a space is to be renovated and a new lighting layout is planned, consider installing new luminaires with state-of-the-art components such as T8 lamps, electronic ballasts and efficient lighting distribution. In existing overlighted spaces, the installation of new luminaires may help gain user acceptance of reduced light levels as occupants focus their attention on the improved quality of the luminaires.

Many lighting upgrades can be performed so that the UL listing on the existing luminaire is maintained. However, some upgrade products are on the market that do not meet UL requirements. By purchasing new UL-listed luminaires, continued UL compliance is assured.

The best choice of a new lighting system would be that which consumes the least input wattage while maintaining the target light level, providing uniform illumination and meeting visual comfort requirements.

Chapter 14

Lighting the Office Environment

T he widespread use of personal computers has caused a dramatic change in the design of office lighting systems. Reduced ambient lighting requirements have rendered the four-lamp fluorescent troffer essentially obsolete. Perhaps the most significant changes in office lighting are the growing use of workstation task lighting and the growing use of direct/indirect lighting systems. In conjunction with other chapters in this book, this chapter provides guidelines for modernizing existing office lighting systems.

OFFICE LIGHTING UPGRADE GOALS

When considering an office lighting upgrade, ensure that the final solution meets the goals listed below:

Visual Comfort
IES RP-1 provides guidelines for visually comfortable lighting in offices. As shown in Table 14-1, RP-1 sets minimum criteria for luminaire brightness at various viewing angles. Choose luminaires that meet the *preferred* criteria where computers are in *constant* use; apply the *minimum* criteria for spaces where computer use is *intermittent*. Where computers are in use, the VCP rating provided for the luminaires should be at least 80 (preferably 90+) in open-plan office areas, and at least 70 in private offices (where there are fewer luminaires that could cause glare). Note, however, that as computer monitor technology improves, these guidelines become less important.

Flexibility
People change. Tasks change. Rooms change. It is difficult to design a static lighting system to meet a dynamic set of conditions. Fortunately,

only one lighting design variable seems to vary significantly—illumination level. Therefore, consider the various ways to design light-output flexibility when upgrading the lighting system. Although manual dimming controls are the most dynamic, they are also the most expensive. Alternatives include step-dimming ballasts, task lights and higher or lower-output lamps.

Life-cycle Cost

With today's selection of highly efficient triphosphor lamps, electronic ballasts and compact fluorescent task lights, high-quality office illumination can be provided with less than one watt per sq.ft. Because energy expenses represent 75-90 percent of a lighting systems' life-cycle cost, maximizing efficiency can also maximize profit.

Table 14-1. IESNA recommendations for maximum luminaire brightness for lighting offices where computers are used. *Source: Illuminating Engineering Society of North America.*

Recommended (Preferred) Criteria	Minimum Criteria
850 cd/m² @ 55° from vertical	850 cd/m² @ 65° from vertical
350 cd/m² @ 65° from vertical	350 cd/m² @ 75° from vertical
175 cd/m² @ 75° from vertical	175 cd/m² @ 85° from vertical

Note: Values shown are expressed in candelas per square meter (cd/m²); these values describe the fixture's relative brightness or luminous intensity when viewed at the specified angles.

RECOMMENDED LIGHT LEVELS

Many offices may be overlighted based on the current recommendations of the IESNA. Refer to Table 14-2 for the recommended light levels for typical office spaces. Note that this table includes illuminance values for both horizontal and vertical surfaces. Chapter 1 provides further guidance in selecting target light levels based on these factors.

Table 14-2. Recommended light levels for office spaces. *Source: Illuminating Engineering Society of North America.*

Activity/Area	Horicontal Footcandles	Vertical Footcandles
Open Plan: Intensive VDT use	30 fc	5 fc
Open Plan: Intermittent VDT use	50 fc	5 fc
Private Office	50 fc	5 fc
Lobby/Reception	10 fc	3 fc
Copy Room	10 fc	3 fc
Mail Sorting	50 fc	3 fc

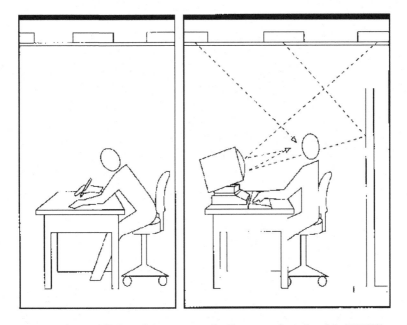

Figure 14-1. As traditional "paper tasks" are replaced with VDT-based tasks, the need for glare shielding becomes critical. *Courtesy: National Lighting Bureau.*

AMBIENT LIGHTING SYSTEMS

Because most offices have relatively low ceilings (less than 20 ft.), most ambient office lighting systems are fluorescent. Regardless of the type of fluorescent luminaires used, triphosphor lamps are recommended for improved efficacy, color rendering and lumen maintenance. The challenge of upgrading the ambient lighting system is to produce low-glare illumination without trapping significant quantities of light within the luminaire.

Downlighting Luminaires
Most office lighting is provided by recessed lay-in troffers which provide 100 percent downlighting. Also known as direct luminaires, downlighting luminaires can be retrofit or replaced with new fixtures to provide high efficiency and glare control.

Deep-cell Parabolic Louvers
A good combination of efficiency and visual comfort is provided by deep-cell parabolic louvers, with no more than 32 cells per 2x4 luminaire (or 16 cells per 2x2). New deep-cell parabolic luminaires with an efficient parabolic reflector/louver design can have a *coefficient of utilization* that is comparable to most lensed fixtures. This means that the light levels produced by high-performance deep-cell parabolic fixtures will be comparable to those produced by lensed fixtures, using the same number

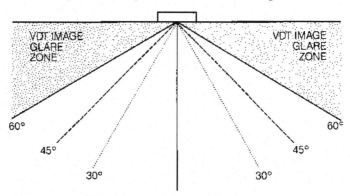

Figure 14-2. When light is emitted at angles above 60° from vertical, reflections in VDT screens are more likely to be seen by workers, which can reduce productivity. *Courtesy: National Lighting Bureau.*

Standard Lensed Luminaire

Full-Chamber Parabolic Luminaire

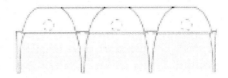

- Efficiency = 70%

- Coefficient of Utilization = 0.74

- VCP = 60

- Efficiency = 66%

- Coefficient of Utilization = 0.72

- VCP = 98

Assumptions: 80/50/20 reflectance, RCR= 1

Figure 14-3. "Full-chamber" parabolic luminaires deliver a comparable coefficient of utilization while vastly improving visual comfort when compared to standard lensed luminaires. *Courtesy: EPA Green Lights.*

and type of lamps and ballasts.

Although lensed fixtures are more efficient, much of their high-angle light output is absorbed by walls and never reaches the workplane. Of course, the disadvantage of deep-cell parabolic luminaires is that the high shielding angle can produce shadows on the upper portions of walls, creating the "cave effect."

Low-glare Clear Lenses

As an alternative to deep-cell parabolic louvers, consider installing a retrofit lens that is specifically designed to reduce high-angle brightness, while maintaining high luminaire efficiency. This lens is less than 1/4-inch thick and can be used for replacing any existing flat lens or diffuser. For most room geometries, the lens will produce a visual comfort probability in the 80s, meeting IESNA recommendations for visual comfort in offices with computers.

Uplighting Luminaires

Uplighting luminaires are either suspended from the ceiling or mounted to systems furniture. By directing some or all of the produced light upward to the ceiling, the brightness of the luminaire is reduced and the appearance of the room is enlarged. Although there is a small sacrifice in the coefficient of utilization, these systems help eliminate the "cave effect." Uplighting systems are particularly useful for reducing shadows from systems-furniture partitions and cabinets. By improving the brightness of walls and ceilings, visual comfort will be improved, since the contrast in brightness within the occupant's field of view is reduced.

TASK LIGHTING SYSTEMS

Task lighting provides illumination where it is most needed—on the work surface—more economically than the most energy-efficient ceiling luminaires, simply because task lighting is located closer to the work surface. By reducing the general illuminance provided by the ceiling fixtures, significant energy savings and improved visibility can result. The task/ambient lighting concept—properly designed and applied—optimizes the office lighting environment by providing ergonomically correct high-quality lighting with minimum power consumption.

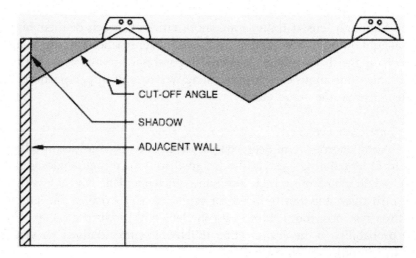

Figure 14-4. Louvered luminaires with a sharp cutoff reduce glare on VDT screens, but they can also produce shadows on the upper portion of adjacent walls. This effect can be reduced by moving the luminaires closer to the wall, installing asymmetric louvers in perimeter locations, installing perimeter uplighting or installing accent lighting. *Courtesy: Illuminating Engineering Society of North America, New York City.*

Linear Undershelf Task Lighting Systems

Systems furniture manufacturers offer a line of linear fluorescent undershelf task lights to compensate for the shadows created by the high partitions and overhanging shelves. Most linear systems use lamp lengths of 3 or 4 ft., shielded by either an acrylic lens or nothing at all. Because most linear task lighting systems are mounted below eye level, direct glare is not a concern. But *reflected glare* can be a concern, particularly when the light source is mounted above and in front of the visual task.

To deal with reflected glare, some linear task lights are produced with "batwing" lenses, multilevel switching, dimming, or light shields. However, some of these measures can reduce luminaire efficiency.

Linear undershelf task lights can be permanently installed in systems furniture, minimizing the possibility of theft. In addition, they do provide some "wall-washing" for illuminating the partition surfaces under shelves, commonly used for bulletin boards.

Compact Fluorescent Task Lighting Systems

Compact fluorescent task lighting systems are more flexible than fixed linear systems, because they can be repositioned for illuminating specific tasks, and they can be aimed by the user. In addition, because the compact fluorescent task lights are mounted to the side of the task, veiling reflections (or reflected glare) are minimized.

Because compact fluorescent task lights produce an essentially circular lighting pattern, many larger workstations may require at least two task lights to properly illuminate reading and writing task areas. Even so, the energy requirements of two compact fluorescent task lights may be less than 25 percent of the energy used by linear fluorescent task lights in a large, 8 ft. x 9 ft. cubicle. Because compact fluorescent task lights can cost from $30 to over $200, the initial investment may be the predominate factor that influences the cost-effectiveness of the retrofit. Other key factors that affect the cost-effectiveness of a task/ambient lighting retrofit are local energy prices and workstation density (number of task lights needed per 1,000 sq.ft.).

Compact fluorescent task lights can be retrofit into existing systems furniture. Some task lights feature a flexible or pivoting lateral support arm which allows users to adjust the vertical and horizontal position of the light. Task lights without articulating arms may be either fixed in position or relocated. The following mounting configurations are available.

Undershelf Task Lights

These task lights are mounted under systems furniture shelves, bins, or cabinets using conventional hardware, magnetic mounting pads or sliding tracks.

Clamp-Mounted Task Lights

This mounting option features a clamp that typically attaches to the desktop or to the systems furniture panel. Clamp-mounted task lights can be used in private offices without systems furniture. Common to each clamp-mounted task light is the articulating arm which enables them to be positioned by the user both vertically and horizontally.

Free-Standing Task Lights

Using a weighted base for support, free-standing desk lamps can be used in virtually any office environment. A wide selection of "executive" style task lights are available that combine a distinctive design with the energy-efficiency of compact fluorescent lamps.

Figure 14-5. Furniture-mounted uplighting and task lighting work together to provide high-quality, energy-efficient illumination. *Courtesy: Peerless Lighting Corporation. Photo by John Sutton.*

Panel-Mounted Task Lights

Task lights can be mounted directly to the systems furniture support structure using manufacturer-specific mounting hardware. Horizontal adjustments are possible with mounting bars that span the distance between systems furniture's vertical supports.

Floor Lamps

Floor lamps feature weighted bases for support, similar to the freestanding desk lamps. However, they do not occupy space on desk surfaces.

Wall-Mounted Task Lights

Using conventional hardware (screws), wall-mounted task lights are mounted on an adjacent wall, using hinged brackets for flexibility in task light positioning.

Figure 14-6. Compact fluorescent task lights can either be free-standing or furniture-mounted in a variety of configurations. *Courtesy: Pritchett Wilson Group, Inc.*

WORKSPACE-SPECIFIC LIGHTING

As a welcome alternative to conventional office lighting systems where luminaires are uniformly spaced, consider the advantages of specifically locating the luminaires over individual workspaces. As shown below, workspace-specific lighting separately distributes light *upwards* for uniform indirect ambient lighting and *downwards* for occupant-specific cubicle task lighting.

Because the uplighting is used to light the circulation areas between cubicles, fewer fixtures are required. Based on a typical 8'x8' cubicle layout, the connected load for office lighting can be reduced by well over 35 percent, yielding power densities of less than 0.7 Watts per square foot. In addition to reducing the connected load, this layout alternative offers additional opportunities for *workspace-specific* occupancy sensing, daylight compensation and personal dimming, which minimize energy consumption while improving user satisfaction.

By suspending direct-indirect luminaires over workstations, fewer fixtures are required and personal control options become feasible. *Courtesy: Ledalite Architectural Products*

Reader Task Lights

Reader task lights can be mounted to a reader's stand (vertical document holder). Some new reader's stands feature built-in task lighting.

LIGHTING CONTROL STRATEGIES

Every lighting upgrade should include an automatic control strategy to eliminate unnecessary lighting operation. The following are suggestions for automatically controlling office lighting systems. For detailed information about automatic switching and dimming systems, see Chapters 11 and 12.

Occupancy Sensors

Because many office workers occupy their offices according to a somewhat irregular schedule, occupancy sensors may be ideal for minimizing the lighting waste that occurs when empty offices are illuminated. Wall-mounted sensors are simple retrofits for small offices; ceiling-mounted sensors may be required in larger offices or where physical constraints limit the effectiveness of a wall-switch occupancy sensor. Where relatively large workstation electrical loads exist (color monitor, coffee warmer, portable heater, radio, etc.), the use of a workstation occupancy sensor may be cost-effective.

Scheduling Controls

Where occupancy schedules are more predictable, scheduling controls can be used. Regardless of the application, all scheduling controls used for office lighting should be equipped with a convenient means for overriding the schedule within a particular zone.

Dimming Controls

Dimming controls in office-space retrofit applications can be difficult to justify on the basis of energy savings alone. Other factors, including improved functionality (e.g., boardroom scene controls) or ergonomics (e.g., personal dimming) should be considered in the investment decision. For example, to justify a personal dimming system in an open office environment, the benefits from dimming energy savings need to be combined with a conservative estimate of improved worker productivity. To further add to the cost-effectiveness, systems that combine personal dimming with occupancy sensing will yield even greater savings.

Figure 14-7. Suspended direct-indirect luminaires with integrated sensors can maximize energy savings in office applications. This integrated sensor provides both occupancy sensing as well as wireless personal dimming. The sensor can be configured to control only the downlight in cubicle spaces; in private offices, the sensor would control both the uplight and the downlight. The sensor is also compatible with DALI communications. *Courtesy: Lightolier*

STUDY CONFIRMS PRODUCTIVITY BENEFITS FROM PERSONAL DIMMING CONTROL

Research conducted by an industry consortium has confirmed that there is a direct connection between the use of personal dimming control and office worker productivity. The Light Right Consortium conducted the research study in an office space in Albany, New York, which was furnished as a typical open plan workplace for nine workers. Two experiments were conducted with a total of six different lighting conditions. The multi-year study found that the presence of personal control had a measurable impact on the motivation of office workers to perform their tasks. Compared to the base case without personal dimming control, subjects with personal dimming control were motivated to sustain their performance as they persisted longer on difficult tasks and were more accurate on a task requiring sustained attention.

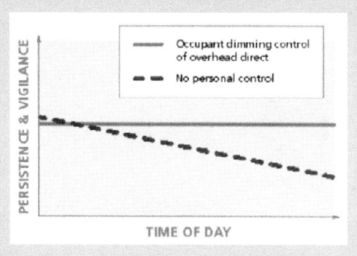

Occupants with dimming control had increased motivation and were able to sustain their persistence and vigilance over time, as compared to those without any control of the lighting. *Courtesy: Light Right Consortium*

Chapter 15

Lighting the Retail Environment

L ighting systems play a vital role in retail applications. Lighting not only allows customers to see the products, it can influence sales. Without effective lighting, the merchandise may not gain attention or arouse interest. However, effective lighting can make products appear more attractive and stand a better chance of being purchased. Properly applied, new lighting upgrade technologies can help improve profits through increased sales and reduced operating costs.

RETAIL LIGHTING UPGRADE GOALS

When upgrading retail lighting systems, consider the lighting upgrade goals presented below. These goals apply in virtually all retail lighting applications.

Product Illumination
The primary objective of retail lighting is to put the merchandise in the "best light." To gain buyer attention and interest, the product must be protected from veiling shadows or reduced illumination. For example, where products are displayed on shelves, the lighting system must efficiently deliver footcandles to *vertical* surfaces such as price tags, product information and the merchandise itself. Vertical illumination can be enhanced by using diffuse fluorescent or coated metal halide sources in luminaires that provide partial uplighting. Alternatively, shelf and display lighting can produce "eye-catching" product brightness.

Communicate Merchandising Strategy
Yes, the lighting system can help communicate the retailer's merchandising strategy. Although most consumers do not realize it, the

lighting system conveys messages regarding pricing strategies, target clientele and even the level of customer service!

Color Rendering

With the availability of triphosphor fluorescent lamps, high-CRI metal halide lamps and a wide variety of halogen lamps, excellent color rendering can and should be incorporated in all retail lighting upgrades. Maximum color rendering is essential in retail applications where decisions are based on discerning subtle variances in color—such as clothing, home-decorating or furniture stores. Although color rendering is less important in other retail applications such as hardware stores, metal halide and fluorescent sources in these applications should have a CRI rating of at least 70.

Flexibility

In some specialty enterprises, the lighting system must evolve with the latest trends. Where feasible, design elements of flexibility into the system in order to minimize future renovation costs.

RECOMMENDED LIGHT LEVELS

The most distinguishing factor that influences retail light levels is the expected traffic volume or activity level. As shown in Table 15-1, the IESNA recommends higher target light levels for higher levels of activity.

Definitions of Retail Activity Levels

IESNA defines three levels of retail activity for which specific illuminance targets are recommended. These activity levels are described below.

High Activity

Common high-activity retail applications include grocery stores, drug stores and discount department stores. In these types of stores, the merchandise displayed has readily recognizable usage, and minimal sales assistance is provided. When selecting products, the customer's viewing and evaluation time is rapid, and merchandise may be shown in such a way to attract and stimulate the impulse-buying decision.

Table 15-1. Target light levels for retail lighting. *Source: Illuminating Engineering Society of North America.*

	Circulation Areas	Merchandise	Feature Displays
High Activity	30 fc	100 fc	500 fc
Medium Activity	20 fc	75 fc	300 fc
Low Activity	10 fc	30 fc	150 fc

Medium Activity

A common example of a medium-activity retail establishment is a large department store. Here, the merchandise is familiar in type or usage, but the customer may require time and/or help in evaluating the merchandise.

Low Activity

These stores include fine jewelry stores, where the type of merchandise is purchased less frequently by the customer. Low activity stores provide plenty of sales assistance to every customer, especially those who may be unfamiliar with the inherent quality, design, value or usage of the product.

HIGH-ACTIVITY RETAIL LIGHTING UPGRADES

First Impressions

In high-activity stores, the marketing strategy is high-volume sales through low prices, as in supermarkets, discount department stores and drug stores. The illumination is very uniform, with very few highlighted feature displays. And the lighting system must have a simple look, suggesting that the money saved on the lighting system translates to "low-low prices." Customers entering these stores will expect a minimal level of customer assistance.

Light Sources

Fluorescent lighting systems are preferred in high-activity stores, mainly because of their low first cost, ease of maintenance and high efficiency. In addition, fluorescent systems create a high level of ceiling brightness that invites customers to enter the store. The most common light sources used in high-activity stores are 8 ft. slimline or high-output fluorescent lamps. Upgrades should utilize triphosphor lamps with electronic ballasts to maintain the relatively high light levels required in these stores. However, as more "discount-warehouse" retail establishments enter the market, energy-efficient metal halide or high-output compact fluorescent systems are becoming popular.

Luminaire Upgrades

In stores with high ceilings (>18 ft.), either strip fluorescent or high-bay clear prismatic luminaires should be used. These luminaires will provide bright ceilings and a high degree of vertical illumination.

Reflector manufacturers have designed retrofit reflectors specifically for improving the coefficient of utilization of strip fluorescent luminaires in retail applications. Arrange a trial installation to evaluate the effects of the reflector on ceiling brightness and vertical footcandles.

With lower ceilings, the effects of direct glare become more important. In such cases, recessed lensed troffers (typically 2' × 4') should be used because they have a lower surface brightness than strip fluorescent fixtures.

MEDIUM-ACTIVITY RETAIL LIGHTING UPGRADES

First Impressions

Medium-activity stores involve a combination of self-service and clerk-assisted product selection activities. In addition to finding more sales assistants, there are more feature displays in these stores compared to high-activity stores. Prices are generally higher than in high-activity stores because of the added time spent with customers. The casual observer will note that the decor in a medium-activity store—such as a department store—is designed to create a more pleasing and comfortable environment, inviting the customer to linger long enough to make the decision to purchase.

Light Sources

Fluorescent equipment is typically used for general illumination, while halogen/incandescent sources are used for display lighting. Fluo-

Figure 15-1. High-activity stores maintain relatively high light levels with few (if any) feature displays. *Courtesy: Lithonia Lighting.*

rescent lighting sources should be upgraded to triphosphor fluorescent lamps and electronic ballasts for maximum efficacy. To compliment the decor and maximize efficacy, fluorescent lamps should have a CRI rating of about 85. Warmer colors are preferred over cool colors as they provide a better color match with halogen display lighting.

To attract attention to featured product displays, illuminate them to levels that are 4-5 times higher than the ambient level. Given the projection distance (the distance between the display lights and the feature display), select halogen PAR lamps with the beam angle that will minimize stray lighting. Good display lighting will restrict the high-intensity illumination to the merchandise itself. This highlighting technique creates a dramatic visual impact, generating attention and interest in the merchandise. Use the formula shown below to calculate the illumination on a display:

$$footcandles = \frac{center\ beam\ candlepower}{(projection\ distance)^2}$$

- Center beam candlepower can be found in lamp catalogs

- Projection distance is measured in feet from the lamp to the display

Luminaire Upgrades

Recessed fluorescent troffers are the norm for medium-activity stores. Where reduced ceiling brightness is desired, deep-cell parabolic louvers should be used. In some medium-activity stores, the best luminaire choice may be a deep-cell parabolic luminaire with a white-enamel finish. These luminaires provide reduced glare compared with lensed luminaires, and they give the store a brighter appearance than specular (or semi-specular) louver surfaces.

Decorative halogen and CFL display lights can be used to match the decor of the store and communicate a level of sophistication to the buyer. Although line-voltage halogen systems will meet many accent

Figure 15-2. Medium-activity stores typically combine low-glare fluorescent ambient lighting with brightly illuminated feature displays. *Courtesy: Lightolier.*

lighting requirements, low-voltage systems may be preferred in applications requiring superior light beam control.

LOW-ACTIVITY RETAIL LIGHTING UPGRADES

First Impressions

Upon entering a low-activity store, customers notice a heightened level of sophistication, quality and expense. There is always a sales specialist ready to assist customers in making their selections. And sales specialists are usually needed for gaining access to the merchandise— particularly in jewelry stores! Because the purchase decisions can be somewhat time-consuming, the lighting system is designed to make the buyer feel comfortable. Compared with high-activity stores, these stores will feature lower light levels, warmer colors, higher color rendering sources, more decorative light fixtures and more accent lighting.

Light Sources

Halogen sources are preferred in low-activity stores for providing both general illumination and display lighting. Point sources, such as halogen PAR lamps or low-voltage halogen MR lamps provide the sparkle and elegance that are commensurate with the store's image.

Luminaire Upgrades

Most luminaires in low-activity stores are either downlights or track lights, both of which could house halogen lamps. Although the emphasis in these stores is quality (not efficiency), decorative compact fluorescent sconces and wall washers should be considered where applicable. With increased visual comfort requirements in low-activity stores, any use of fluorescent lighting should utilize low-brightness louvers, not lenses.

LIGHTING CONTROL STRATEGIES

Because of their regular operating hours, retail lighting systems should be controlled using some form of scheduling system. Small boutiques typically utilize a simple electronic time clock. Larger department stores may employ a time-based dimming and switching system that can

Figure 15-3. Display lighting should be illuminated to levels that are 4-5 times higher than the general ambient illumination. *Courtesy: Philips Lighting Company.*

schedule changes in light levels as stocking and maintenance operations conclude and merchandise shopping begins.

Another form of dimming control could be applied in 24-hour grocery stores that normally provide high light levels for rapid product identification and evaluation. However, at night, the store's ambient lighting can be dimmed to save energy without reducing the visibility of the product. How is this possible? During bright sunny days, the lights must be at a higher level to allow people who enter the store from the bright outdoors to see in the darker enclosed space of the store. Conversely, lighting power can be reduced substantially at night, as human eyes are "night-adapted" and do not require as much light for the same visual performance, saving as much as 50 percent or more during nighttime hours. This dimming application is known as adaptation compensation.

Figure 15-4. Low-activity stores provide maximum glare control and reduced levels of general illumination. Dramatic display lighting is achieved with halogen lamps that provide precise beam control. *Courtesy: Philips Lighting Company.*

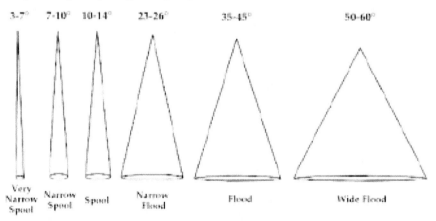

Figure 15-5. Directional lamps for display lighting are available in a wide range of beam angles. The choice of beam angle depends on the size of the target to be illuminated and the distance of light projection. *Courtesy: EPA Green Lights.*

Chapter 16

Lighting the Industrial Environment

Since the dawn of the industrial revolution, industrial lighting has been designed to promote safety and productivity. However, in recent years energy efficiency has become another key motivator. Until the early 1970s, energy costs for industrial customers were low, frequently less than $0.01 per kWh. The escalation of energy prices that began in the mid-1970s has sparked new lighting system efficiencies.

INDUSTRIAL LIGHTING UPGRADE GOALS

Although illumination requirements within industrial facilities vary widely, all lighting systems in these facilities should contribute to safety, productivity and energy efficiency. To achieve these goals, the lighting system design must meet the following objectives:

- Provide the proper ambient and task-specific light levels.

- Utilize appropriate light sources and luminaires based on dirt conditions, room dimensions and task orientations.

- Operate effectively under potentially harsh conditions.

- Minimize energy consumption while meeting needs for discerning color.

- Eliminate the potential distraction or hazards of stroboscopic lighting near rotating machinery.

RECOMMENDED LIGHT LEVELS

There are literally hundreds of unique visual tasks that take place in industrial settings. The industrial lighting system should be designed to deliver the illumination needed to perform the specific tasks planned for that area of the facility. The proper quantity of illumination—measured in footcandles—is dependent on the visual requirements of the tasks performed as well as the visual capabilities of the occupants. Note that as industrial processes have changed over the past decade, the existing lighting systems may need to be modified to provide a new level of illuminance and quality.

Table 16-1 lists a few of the many types of visual tasks performed in industrial facilities along with their recommended illuminance values. To identify the recommended illuminance values for other industrial visual tasks, see the IESNA *Lighting Handbook*.

Table 16-1. IESNA illuminance values for selected industrial activities. *Source: Illuminating Engineering Society of North America*

Area/Activity	Recommended Illuminance
Component Manufacturing	
Large	30 fc
Medium	50 fc
Fine	100 fc
Raw Material Processing	
Coarse	10 fc
Medium	30 fc
Fine	50 fc
Assembly	
Simple	30 fc
Difficult	100 fc
Exacting	300-1000 fc
Warehousing	
Inactive	5 fc
Active: bulky items	10 fc
Active: small items	30 fc

OPTIONS FOR LIGHTING SYSTEM LAYOUT

Up to three methods can be used for distributing light in industrial spaces: general illumination, aisle illumination and task-oriented illumination.

General Illumination

General illumination lighting layouts produce a uniform light level on a horizontal workplane using an array of evenly spaced, ceiling-mounted luminaires. For example, general illumination lighting systems may be used for illuminating shipping/receiving areas or bulk storage areas. In some applications, aisles of merchandise may not be in permanent locations, so the lighting system must remain flexible to illuminate shelved material, regardless of location. In such cases, using luminaires that produce a higher percentage of up-lighting will result in a more diffuse quality of light, thereby reducing shadows caused by relocated machinery or stacks.

Aisle Illumination

Aisle illumination consists of luminaires mounted directly over the centerline of each aisle, typically intended to illuminate warehouse shelves. In these aisle-specific lighting applications, the system should achieve the desired light level on vertical surfaces, such as carton labels. Fixtures used to illuminate aisles include either fluorescent or unique HID luminaires that are designed to distribute light in an elongated "aisle shaped" pattern.

Figure 16-1. In warehouse aisles, lighting upgrades should maintain the target illuminance on vertical surfaces. This can be achieved with clear prismatic reflectors. *Courtesy: Lithonia Lighting.*

Task-Oriented Illumination

Tasks that are performed in specific locations may require higher light levels than are provided by the general or aisle-specific illumination systems. The additional footcandles required for these tasks can be provided by one of the following methods:

1.   Increasing the concentration of luminaires and/or increasing the lumen output of the general illumination system over the task locations.

2.   Providing supplemental task lighting in the specific locations where the tasks are performed.

For some applications, such as inspection lighting, the strategic positioning of task lighting can create a grazing light which enhances certain inspection processes. In all task lighting applications, be careful to position the task lighting so that contrast on the visual task is enhanced rather than veiled by reflected glare.

INDUSTRIAL LUMINAIRE SELECTION FACTORS

The choice of a luminaire for a specific industrial application is driven by several factors. These factors include:

•   Luminaire mounting height.
•   Orientation of visual tasks.
•   Dirt conditions.
•   Harsh environmental conditions.

Luminaire Mounting Height

The selection of luminaires in industrial facilities depends on how far the light must be projected. As a general guideline, *high-bay* lighting systems are mounted at least 20 ft. above the floor, while *low-bay* systems are mounted at heights less than 20 ft.

High-bay applications usually require the use of HID sources such as metal halide or high-pressure sodium. These systems are preferred in high-bay applications because they can focus and direct light more efficiently than the more diffuse fluorescent sources. In some applications

Figure 16-2. Task lighting should be used wherever higher light levels are needed for enhancing safety and productivity. Typical applications include machine operation, inspection and assembly. *Courtesy: Illuminating Engineering Society of North America.*

where a more diffuse source is needed for reduced glare, wide distribution or enhancing vertical footcandles, compact fluorescent high-bay luminaires may be specified. However, consider performing a trial installation to verify that the compact fluorescent system will provide acceptable performance under potentially hot ambient conditions. Refer to Chapter 8 for more information about these luminaires.

In applications with mounting heights under 20 ft., the most cost-effective lighting systems are usually fluorescent. In low-bay applications, consider the use of T8 or T12 industrial strip fixtures with electronic ballasts. Where relatively high mounting heights (15-25 ft.) and high light levels (>30 fc) are required, high-output fluorescent lamps and electronic ballasts should be considered.

As an alternative to industrial fluorescent strip fixtures in low-bay applications, consider using low-bay HID luminaires, particularly in areas with low ambient temperatures. These luminaires use a prismatic lens to provide wide-angle light distribution and glare shielding.

Orientation of Visual Tasks

Unlike office environments where the visual tasks are oriented in the horizontal plane, industrial visual tasks may also require illumination on other workplane orientations. For example, tasks such as operating machinery, operating a forklift or inspecting solid objects may require "spherical" illumination (lighting on all workplanes). To achieve spherical illumination, the lighting system is designed to maximize illumination on the vertical

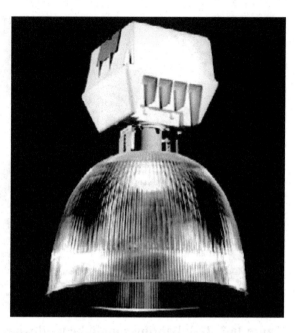

Figure 16-3. In relatively clean environments, this high-bay luminaire can be used to provide uplighting for improving illumination on all visual task orientations. *Courtesy: Thomas Lighting.*

plane, which usually translates to relatively uniform illumination on other workplane orientations.

To enhance illumination on vertical surfaces, the luminaires must be designed to direct a portion of the light upward toward a white ceiling that diffuses the light. The scattering of light results in a higher percentage of lumens striking the vertical plane. To maintain this vertical illumination component from uplighting, the ceiling reflectance should be maintained by periodically painting the ceiling white. In dirtier conditions, either the ceiling will need to be painted more frequently or the uplighting option needs to be abandoned.

Dirt Conditions

Dirt and dust produced by industrial processes accumulate on luminaires, lamps and room surfaces, reducing light levels and efficiency by as much as 50 percent. In order to maintain luminaire efficiency and

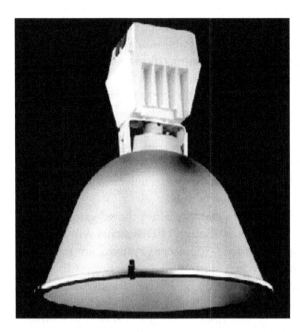

Figure 16-4. When used in dirty environments, HID luminaires should be enclosed and gasketed to keep dirt off the lamps and reflector surfaces. And use enclosed HID luminaires where metal halide lamps are not rated for open luminaire operation. *Courtesy: Thomas Lighting.*

reduce maintenance expenses, luminaires in these environments should be designed to minimize dirt accumulation.

In clean and very clean industrial environments with mounting heights of 20-40 ft. above the floor, HID fixtures with clear prismatic reflectors are recommended. These reflectors efficiently direct the light to the floor while allowing a small percentage of the light to reflect off the ceiling to improve vertical illumination. At lower mounting heights in clean environments, industrial strip luminaires with slotted reflectors can provide needed uplighting and downlighting components.

In moderately dirty environments, open ventilated fixtures will stay cleaner than open-bottom, closed-top units. The ventilated fixtures allow convected air flow which carries dirt up and out of the fixture. However, in areas of heavy or very heavy dirt contamination, consider the use of enclosed, gasketed fixtures which are sealed against dirt entry.

Periodic washing of fixtures results in a higher overall lighting system efficiency. In some cases, initiating a fixture cleaning program can allow for a reduction in luminaires needed to provide the target light level, resulting in significant energy cost savings.

Harsh Environments

Luminaires that are located in hazardous, damp, wet, corrosive, hot or cold environments must be designed to meet specific operating requirements. Some industrial environments—such as paint-spraying booths or textile processing areas—contain explosive or flammable vapors, dust or fibers. These locations require fixtures that meet local building standards and are classified for use in these hazardous environments. Similarly, luminaires located where large amounts of water are used may need to be approved for use in either damp or wet locations.

Corrosive vapors that may be present in some industrial operations can attack and destroy conventional luminaires. All exposed surfaces—including mounting hardware—must be protected by a finish that is capable of withstanding the corrosive action.

Extreme ambient temperatures can also affect the selection of lighting systems in industrial facilities. Although HID systems will start in a wide range of ambient temperatures, fluorescent systems are more sensitive to ambient temperature when starting and operating. If the ambient temperature is expected to be below 50°F when the lighting system is activated, low-temperature ballasts should be specified that are capable of starting lamps at temperatures below 0°F. These low-temperature ballasts are available in either magnetic or electronic versions. Low-temperature electronic and magnetic ballasts for high-output (800 mA) and very high output (1500 mA) lamps can start lamps at -20°F. Refer to Chapter 3 for minimum starting temperatures for various fluorescent systems.

The light output of fluorescent lighting is also affected by temperature. Fluorescent lamps provide relatively high lumen output at room temperatures. However, fluorescent light output can be vastly diminished in ambient temperatures that are above or below room temperature. Installing an insulating tube guard or shield around fluorescent lamps helps improve fluorescent lamp output in cold temperatures. Refer to Chapter 3 for guidance in determining application-specific, temperature-corrected light output and energy consumption data.

LIGHT SOURCE SELECTION FACTORS

Once the luminaires have been selected based on the physical layout and environmental conditions, choose the light source that meets the requirements for discerning color, energy efficiency and visual performance.

Discerning Color

In applications where discerning color is critical, the lighting system must employ lamps with a relatively high CRI rating, such as fluorescent, metal halide, or color-corrected (deluxe) HPS lamps. Where color quality is less critical, standard HPS systems are preferred because of their higher efficacy. Note that recent studies have demonstrated improved visibility in low light levels ($\leq$50fc) when higher color temperature (>4000K) sources are used, such as fluorescent and metal halide.

Efficacy

Table 16-3 lists the common light sources used in industrial applications. In fluorescent systems, the maximum efficacy is achieved with electronic ballasts and triphosphor fluorescent lamps with a CRI rating of 82-85. For maximum HID efficacy, select standard HPS systems. And

Table 16-3. Typical performance of industrial light sources. *Source: Manufacturer literature.*

Light Source	CRI Rating	Maintained Efficacy (lm/W)
FLUORESCENT		
Compact Fluorescent	82-86	60-80
T8 (265mA)	75 - 95	50 - 90
T5 High Output	85	80-85
T12 Standard (430mA)	62 - 90	50 - 90
T12 High Output (800 mA)	62 - 90	62 - 81
T12 Very High Output (1500mA)	62 - 90	42 - 44
HIGH-INTENSITY DISCHARGE		
Metal Halide	65 - 85	36 - 86
High-pressure Sodium		
Standard HPS	22	45 - 115
Deluxe HPS	65	42 - 73
Mercury Vapor	22 - 50	19 - 43

where color rendering is important, use position-specific metal halide lamps. Note that higher efficacy is achieved in higher-wattage HID systems. Therefore, using fewer high-wattage luminaires may yield significant energy savings.

HPS Stroboscopic Effect

HPS lamps tend to strobe and may cause a reduction in visual performance (or even a safety hazard) if used with rotating machinery. Where *single-phase* power distribution systems exist, supplement the HPS lighting with electronically ballasted fluorescent task lighting in order to minimize the stroboscopic effect. In *three-phase* power distribution systems where rotating machinery is present, adjacent luminaires should be staggered across the three phases so that the HPS lamps will not be strobing synchronously. Other lamp types, including metal halide, do not normally create strobe problems. Special retrofit metal halide lamps can be used on existing HPS ballasts to improve color rendering and minimize strobe effects, but they can reduce illumination by 25 percent or more.

INDUSTRIAL LIGHTING UPGRADE STRATEGIES

Put simply, the approach for maximizing savings in industrial lighting systems is to determine the task-specific light level requirements for each area in the plant or warehouse, and to use the most efficient systems for meeting those requirements. Then, make sure that the lighting systems are turned off at the end of the last shift with the use of automatic scheduling controls.

Upgrade To More Efficacious Systems

Using systems that achieve target light levels using the least wattage is one of the key strategies for upgrading industrial systems. Table 16-4 provides a rough guideline for upgrading high-bay mercury vapor systems with more efficacious sources. For detailed information about fluorescent and HID lighting upgrade options that may apply to specific industrial applications under consideration, refer to Chapters 4 and 8, respectively.

Task-Specific Light Levels

The existing light levels may be reduced if most of the tasks in a given area do not require high levels of visual acuity. In such cases, the

Table 16-4. HID upgrade options for high-bay mercury vapor systems.
Source: Manufacturer literature.

Existing System	Retrofit Option	Reduction In System Wattage	Relative Light Output, Maintained
175W Mercury	70W HPS	111	74%
Vapor	70W MH/Ceramic	116	67%
	70W MH/Standard	116	54%
	100W HPS	76	116%
	100W MH/Ceramic	79	103%
	100W MH/Standard	79	86%
250W Mercury	150W HPS	89	137%
Vapor	150W MH	89	97%
	175W MH	74	114%
400W Mercury	150W HPS	263	82%
Vapor	250W HPS	158	154%
	250W MH	165	103%
	325W Retrofit MH	75	104%
1000W Mercury	250W HPS	750	60%
Vapor	400W HPS	585	100%
	400W MH	592	71%
	950W Retrofit MH	20	178%

Notes: Deluxe HPS performance is similar to metal halide; MH = metal halide; HPS = High Pressure Sodium.

visual tasks that do require much higher light levels can be performed with the use of supplemental task lights. If the visual tasks in a given area are expected to vary over time, consider installing systems that can be step-dimmed or continuously dimmed based on the task's visual requirements.

Reduce Fixture Quantities

In some applications, the least-wattage lighting solution may involve the use of fewer, *higher-wattage* luminaires that are inherently more efficient. For example, a 400W HPS system produces about 100 maintained lm/W, compared to a 150W HPS system which produces about 74 maintained lm/W. The same space may be illuminated with one-third of the luminaires if 400W HPS lamps are used instead of 150W systems. However, when reducing the number of luminaires in a space, make sure that illumination uniformity is not sacrificed by exceeding the new luminaires' spacing criteria. Although most upgrades will use existing socket locations, there are situations where it may be economical to relocate fixtures to optimal locations for maximum energy savings and uniform illumination

Increase Luminaire Efficiency

Retrofit reflectors can be added to industrial strip luminaires to increase efficiency. Some retrofit reflectors will include slots that allow some uplighting to increase illumination on vertical surfaces. In many applications, retrofit reflectors can be installed while preserving the combined functions of brightening room surfaces and meeting task illumination needs.

Unlike most 4 ft. reflector applications, many 8 ft. reflector upgrades do not involve delamping. For example, a 2-lamp VHO system could be replaced with a 2-lamp HO system plus a retrofit reflector. Or, a two-lamp HO system could be replaced by a slimline system plus a reflector (and new single-pin sockets). The retrofit reflector material could be a high-gloss powder-coat white paint or a specular (mirror-like) reflector. The choice of reflector material is driven by dirt conditions, cost, the condition of the base system's reflector, and light level requirements. To maintain the increased efficiency of the luminaires, initiate a schedule of routine fixture cleaning, based on the rate of dirt buildup on reflector surfaces.

Retrofit reflectors can also be used to improve HID luminaire efficiency. Either specular or clear reflectors can be used to improve the luminaire output without increasing energy input. To save energy in these applications, HID power reducers or reduced-wattage "energy-saver" HID lamps can be used. Trial installations are strongly recommended with these products to determine the effects on light output and distribution.

Automatic Controls

Additional energy savings in industrial lighting can be achieved with the use of automatic scheduling or dimming controls. A scheduling control system would be used to automatically turn off the lights after

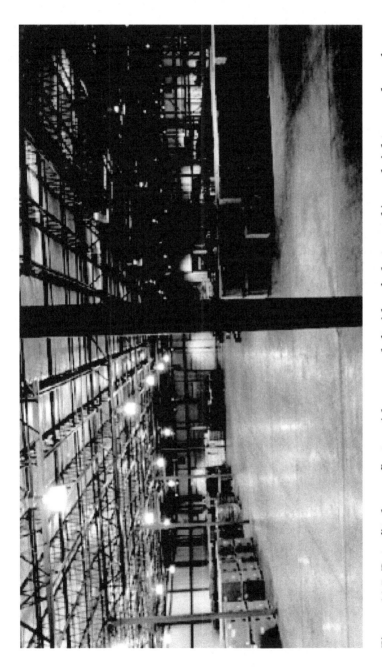

Figure 16-5. Retrofit clear reflectors (shown on left side of photograph) can brighten work surfaces while maximizing luminaire efficiency. *Courtesy: Lexalite International.*

the last operating shift. In applications with skylights, the ambient lighting system could be deactivated or dimmed in response to the daylight levels detected by photosensors. To dim a circuit of HID luminaires, consider installing a panel-level dimming system which reduces the power to the HID circuit based on manual, daylighting or schedule inputs. Alternatively, step-dimming HID or compact fluorescent systems may be used.

The use of occupancy sensors may be used to turn off or dim aisle-specific lighting systems when warehouse aisles are unoccupied. Because of their "instant-on" characteristic, fluorescent systems are more suitable for control strategies involving frequent on-off operations. HID systems, however, require a few minutes to "restrike" and warm up after being turned off. Therefore, occupancy sensing for HID systems should incorporate bi-level switching (using capacitive switching ballasts). Alternatively, instant-restrike HPS lamps may be used, provided that the HPS warm-up delay is acceptable. For more information about applications of automatic switching and dimming controls, see Chapters 11 and 12.

Chapter 17

Outdoor Lighting

A fter the sun sets, life goes on. Shipping, maintenance and materials handling operations often continue in outdoor locations long into the night, and effective outdoor lighting can enhance the productivity of these activities. As daylight vanishes, the potential for accidents and crime increases. Therefore, conservation efforts directed toward outdoor lighting systems should not sacrifice productivity or safety.

OUTDOOR LIGHTING UPGRADE GOALS

Every outdoor lighting upgrade should satisfy each of these goals:

- Deliver the light level needed to provide adequate security, safety and productivity.

- Minimize visual fatigue by maintaining relatively uniform light levels and controlling glare, particularly in applications involving vehicle use. The human eye is easily fatigued when it must continually adapt to differing degrees of brightness. In some cases, direct glare can severely reduce visibility, increasing the risk of accidents.

- Select light sources that provide the appropriate color rendering performance. In activities such as materials handling and outdoor retail activities, color identification can be critical. However, a high CRI rating is not important for routine surveillance where the visual task is detecting potential security problems.

- Use the light sources and luminaires that will most economically deliver lumens to the outdoor area. This may involve upgrading to

263

a more efficient outdoor lighting source and/or installing efficient luminaires with improved lighting distribution performance.

• Install automatic controls to eliminate daytime operation and minimize unnecessary nighttime operation.

RECOMMENDED LIGHT LEVELS AND UNIFORMITY RATIOS

The IESNA recommends specific maintained illuminance values for outdoor lighting applications. Table 17-1 lists these recommendations for some of the more common applications. Note that where vehicle operation is assumed, the lighting system should maintain a uniformity ratio that does not exceed the maximum value shown. For example, assume that the average maintained light level in a covered parking area is 5 fc. To maintain an average-to-minimum uniformity ratio of no more than 4:1, the minimum light level (typically between luminaire locations) should be no less than 1.25 fc. The IESNA *Lighting Handbook* lists illuminance recommendations for many other outdoor lighting applications not shown in the table.

LIGHT SOURCE SELECTION FACTORS

To maximize energy savings, use the lowest-wattage outdoor lighting systems that deliver the appropriate light levels and desired color quality. The variables of color rendering requirements and mounting height will significantly affect the choice of upgrade solutions. Other factors that affect light source selection are the minimum starting temperature and the physical layout of the outdoor area to be illuminated.

Color Rendering

Under sources with a low CRI, colors will appear unnatural or less bright than under high CRI sources. Therefore, avoid using HPS and LPS sources in outdoor lighting applications where high quality color rendering is important (such as car sales, billboards and sports facilities). For example, metal halide lamps may be used to illuminate a softball field (where color rendering is important), while high-pressure sodium lamps may be used to illuminate the roadway and parking areas. Refer to Table

17-2 for typical CRI and efficacy values of light sources used for outdoor applications.

As Table 17-2 shows, the most efficacious outdoor lighting sources generally deliver the lowest color rendering performance. However, in many outdoor applications, lighting system selections are driven by color rendering and mounting height factors, leaving relatively few decisions to make regarding efficacy. To the extent possible, avoid using

Table 17-1. Target light levels and uniformity ratios. *Source: Illuminating Engineering Society of North America.*

Outdoor Lighting Applications	Recommended Maintained Illuminance (And Maximum Uniformity Ratios)	
Building Exteriors	5 fc	entrances
	1 fc	surroundings
Billboards/Signs	100 fc	dark surfaces; bright surroundings
	50 fc	light surfaces; bright surroundings
	50 fc	dark surfaces; dark surroundings
	20 fc	light surfaces; dark surroundings
Gardens	0.5 fc	general lighting
	1 fc	path and steps
Loading Platforms	20 fc	
Recreational Sports	10 fc	basketball
	10 fc	tennis
	10 fc	softball (infield)
	7 fc	softball (outfield)
Local Commercial Roadways	8 fc	(6:1 average-to-minimum uniformity ratio)
Covered Parking Facilities	5 fc	(4:1 average-to-minimum uniformity ratio)
Open Parking Facilities: General Parking & Pedestrian Area	3.6 fc	high activity (4:1 avg/min uniformity)
	2.4 fc	medium activity (4:1 avg/min uniformity)
	0.8 fc	low activity (4:1 avg/min uniformity)
Open Parking Facilities: Vehicle Use Area (only)	2.0 fc	high activity (3:1 avg/min uniformity)
	1.0 fc	medium activity (3:1 avg/min uniformity)
	0.5 fc	low activity (4:1 avg/min uniformity)

Table 17-2. Typical performance of outdoor light sources. *Source: Manufacturer literature.*

Light Source	CRI Rating	Maintained Efficacy (lm/W)
FLUORESCENT		
T4 Compact Fluorescent	82 - 86	25 - 65
T5 Twin-Tube Fluorescent	82 - 85	35 - 74
T8 Standard	75 - 85	70 - 90
T12 Standard	62 - 85	50 - 90
T8 High Output	85	81
T12 Very High Output	62 - 85	42 - 44
HIGH-INTENSITY DISCHARGE		
Metal Halide	65 - 85	38-86
High-pressure Sodium		
Standard HPS	22	45 - 115
Deluxe HPS	65	42 - 73
Mercury Vapor	22 - 50	19 - 43
Low-pressure Sodium	0	50 - 150

low-efficacy sources such as incandescent, mercury vapor, VHO fluorescent and other magnetic-ballast fluorescent systems.

Mounting Height

To illuminate large outdoor areas with high-mounted luminaires, use either HPS or metal halide lamps. These are energy-efficient, high-output "point" sources. Point source luminaires are inherently the most effective for long-distance light projection.

For lighting smaller areas and/or where luminaires are mounted at lower heights, the more diffuse sources may be used. Fluorescent and LPS lamps are "low-pressure discharge" sources which, unlike point sources, provide relatively diffuse (scattered) light and are more effective at lower mounting heights (below 20-25 ft.). HID sources may also be used at these lower mounting heights, provided that adequate glare shielding is provided. The mounting height of low-wattage (<18W) compact fluorescent sources should generally be limited to 10 ft. or less.

Cold-weather Performance

Note that the light output and efficacy of fluorescent systems can drop dramatically in temperatures under 50°F, depending on the lamp type and chemistry. Some compact fluorescent lamps utilize a mercury amalgam (mercury alloy) that allows for cold-weather starting and near-peak operating performance at temperature extremes. With low-temperature fluorescent electronic ballasts available for most standard and high-output full-size fluorescent lamps, low-temperature starting and high efficacy can be achieved with outdoor fluorescent lighting systems. Where low temperatures are expected to prevail, clear tubular insulating jackets can help increase lumen output from linear fluorescent systems.

Both low-pressure and high-pressure sodium sources will start reliably at temperatures as low as -40°F. Most metal halide systems require a minimum starting temperature of -20°F. And unlike fluorescent systems, HID system wattage and lumen output are essentially independent of ambient temperature.

OUTDOOR LUMINAIRES

To maximize effectiveness and minimize waste and light trespass, outdoor luminaires should confine the lighting to the target area. The choice of outdoor luminaires is driven by the size of the outdoor area, the need to control spill light, and aesthetics.

High Mounting Luminaires

High mounting luminaires generally use HID sources and are mounted on poles at least 15 ft. high. These luminaires are described below and illustrated in Figure 17-1.

High Mast

With mounting heights of 60-100 ft., high mast luminaires are used to illuminate large areas. These systems are usually equipped with a lowering device for convenient maintenance at ground level. High mast lighting is typically used for highway interchange lighting and off-street areas such as industrial yards and large parking lots.

Refractor

Refractor luminaires provide wide beam distribution and are generally used where spill light control is not important. The wide beam

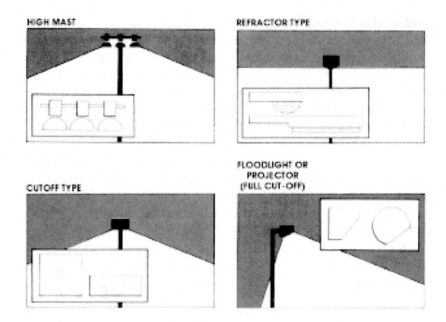

Figure 17-1. Common high-mount outdoor luminaires. *Courtesy: National Lighting Bureau.*

distribution allows for wider spacing of luminaires. Typical applications include highway, street and general area lighting. The refractor luminaires that use fluorescent or LPS lamps need to be mounted at lower heights because of their reduced optical control.

Cut-Off
      Primarily used for lighting medium to large areas, this luminaire is very effective in minimizing direct glare. Where aesthetic appearance is a concern for low-glare street lighting and parking lot applications, a low-profile, horizontal lamp unit may be used.

Floodlight
      Using an efficient reflector for sharp cut-off control, these luminaires are typically used where control of glare and light trespass are critical. Typical examples include airport apron lighting or areas immediately adjacent to residential properties.

Medium and Low Mounting Luminaires

Medium and low mounting luminaires may use compact fluorescent, LPS or HID sources. Mounting configurations for these include both building-mount and pole-mount. These luminaires are described below and illustrated in Figure 17-2.

Building-mounted Refractor

Using lower wattage HID lamps, building-mounted refractor luminaires produce a wide beam distribution for general lighting around buildings. Without a sharp cut-off, however, its glare may be objectionable in some applications.

Building-mounted Cut-off

Instead of using a refractor/lens, this luminaire uses a reflector to achieve tight beam control and low brightness. The upper part of the beam is cut off to prevent light trespass beyond the intended illumination area.

Post-top

Typically used for pedestrian walkways and small-area lighting, these decorative luminaires can utilize low-wattage compact fluorescent or HID sources. Some post-top luminaires feature diffusing globes that allow the light to be distributed in all directions, wasting most of the light that is emitted above horizontal. "Controlled" post-top luminaires are more efficient because they direct the majority of the lamp light onto the desired area.

Low-mounted Site Lighting

Compact fluorescent and HID sources are generally used in low-mounted site

Building-mounted (Refractor Type)

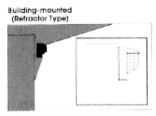

Building-mounted (Cut-off Type)

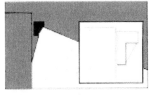

Post-top (Uncontrolled)

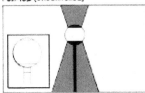

Post-top (Controlled)

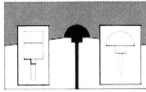

Low-mounted Site Lighting

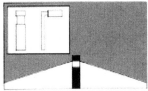

Figure 17-2. Common medium and low mounting outdoor luminaires. *Courtesy: National Lighting Bureau.*

lighting to provide illumination for walkways and small areas. Mounted below eye level, these decorative luminaires can efficiently deliver glare-free light.

OUTDOOR LIGHTING APPLICATIONS

Table 17-3 provides guidance in choosing light sources for outdoor lighting applications. The four applications shown in the table illustrate the diversity that exists in outdoor lighting applications.

Table 17-3. Light sources for outdoor lighting applications.

	High-Pressure Sodium (CRI>22)	Metal Halide (CRI>65)	Linear Fluorescent (CRI>62)	Compact Fluorescent (CRI>80)	Low-Pressure Sodium (CRI=0)
Large Area; High Mounting (Parking Lot)	•	•			
Small Area; Low Mounting (Walkway)	•	•		•	(•)*
Sign/Billboard		•	•	•	
Covered (Parking Garage)	•	•	•		(•)*

*LPS lamp use is limited to applications where monochromatic sources are acceptable.

Parking Lots (Large Areas)

Effective outdoor parking lot illumination can attract customers to retail establishments, promote traffic and pedestrian safety, deter crime and vandalism, and create a sense of personal security. In addition to selecting efficient light sources, energy-efficient parking lot lighting must provide proper light distribution.

Parking lot luminaires should efficiently direct the light to the parking surface. Even if an extremely efficacious light source is used, a narrow lighting distribution will cause uneven illumination (or will require more luminaires in new installations). Therefore, the selection of the luminaire's optical system is part of the lighting upgrade design. Some luminaires that use interchangeable or rotational optical systems allow users to change the lighting distribution after the luminaire is installed.

By limiting the illumination to the parking surface, lighting energy is reduced and neighboring inhabitants won't have the annoyance of light trespass. In general, luminaires that limit light output at high angles (above 75°) reduce the potential for light trespass.

Walkway/Architectural Lighting

Walkway and architectural lighting can be used for facilitating pedestrian safety and traffic while enhancing the outdoor appearance. Architectural luminaires that use HID sources include post-tops and bollards, many of which provide indirect light with reflective surfaces to reduce glare. Lower level illumination applications such as pathway or garden lighting may use low-wattage compact fluorescent or low-voltage halogen sources.

Signs

Effective sign lighting is essential for the sign's message to be communicated at night. Signs can be illuminated internally or externally. The goal of sign lighting is to provide high visibility through the proper selection of light sources.

Internally illuminated signs are used widely in retail applications, typically utilizing linear fluorescent lamps. In these applications, color rendering is not as important as color temperature. Typically, cool-white (4100K) provides good results. EPACT has allowed full-wattage cool white lamps to be used in outdoor sign applications, because the energy-saver (reduced wattage) alternatives require a much higher (60°F) minimum starting temperature.

Externally illuminated signs include most roadway signs and billboards. Light sources with cooler color temperatures (>4000K) are normally chosen for these applications, because they can improve nighttime visual acuity, which is needed for reading words and numbers at a distance. Compared with warmer sources (such as HPS), cooler sources (such as metal halide and fluorescent) cause the pupil to constrict to a

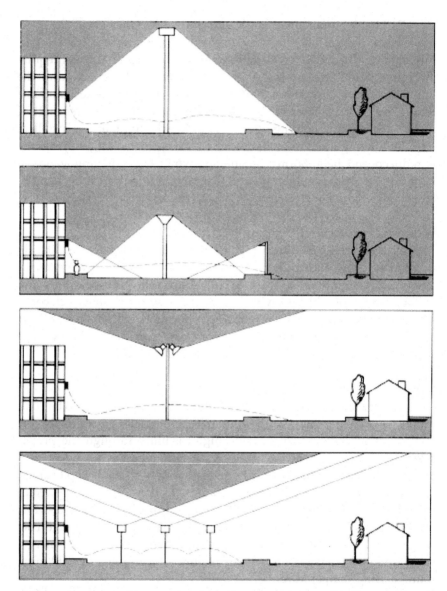

Figure 17-3. Energy can be saved by controlling the distribution of light from outdoor systems. The top two illustrations show how the lighting can be confined to the target area without wasting energy on spill light. However, the bottom two illustrations show that light trespass not only wastes energy, but can aggravate the neighbors. *Courtesy: National Lighting Bureau.*

smaller size, thereby creating a more distinct image on the eye's retina. The specific light source that is chosen depends on the size of the sign and the distance the light source is placed from the sign.

Covered Parking Garages

Many drivers find that operating a vehicle in a parking structure can be extremely challenging. Therefore, improved garage lighting can lead to fewer accidents. The most difficult aspect of garage lighting is providing high uniformity *and* low glare. The worst garage lighting systems force the driver to squint and strain as the car passes between luminaires just before it is parked in a dark stall.

Although both fluorescent and HID sources can be used, most new installations utilize fluorescent sources with low-temperature ballasts. Fluorescent sources provide diffuse illumination, resulting in reduced shadows. Properly shielded, fluorescent systems can provide a low glare garage lighting environment as well. If HID sources are chosen, a sharp cutoff angle can help improve visual comfort, but it reduces uniformity.

LIGHTING CONTROL STRATEGIES

All outdoor lighting systems should be automatically controlled using a daylight switching system. At a minimum, photocells should be used to activate the lighting system at dusk and turn it off at dawn. However, when replacing the older mechanical photocells, consider reducing energy and maintenance costs by installing new solid-state electronic photosensors which combine longer service life with more accurate daylight sensing.

If the outdoor lighting is not needed throughout the night, a timed switching system may be wired in series with the photosensor to switch off the circuit before dawn. Microprocessor-based timed switching systems are an alternative to photosensors. They predict seasonal dusk/dawn switching times and automatically switch the lighting systems accordingly. Microprocessor-based daylight switching systems can incorporate "pre-dawn" scheduled switching.

Bi-level switching is another control strategy for saving outdoor lighting energy. In many applications such as parking lots, recreational tennis courts and security lighting, full light output is not needed unless the space is occupied. Using an infrared occupancy sensor with a bi-level switching system, the light sources become brighter when an occupant is detected. This control strategy not only saves energy, but it can also be effective in deterring crime and vandalism. Security guards will know if a space has been recently occupied if the lights are bright. New HID luminaires are available with bi-level switching capability. Some bi-level HID luminaires are sold with dedicated occupancy sensors; others must receive the occupancy signal from remote sensors via control wiring or a powerline carrier signal.

Chapter 18

Assessing Lighting Upgrade Opportunities

Many attractive investment opportunities available today can be found in the ceiling. Lighting upgrade investments can yield returns that far exceed those of competing low-risk investments. Beyond the obvious cash flow advantages of lighting upgrades, other benefits include improving worker productivity. But before the system is selected, several steps must be taken to verify that the upgrade will yield the desired results.

OVERVIEW

With proper planning and evaluation, lighting upgrades will yield rapid returns to the organization, while helping to reduce the waste and pollution caused by inefficient lighting. This chapter provides guidance in each step of the lighting upgrade assessment phase. These steps are listed below:

- Prioritizing facility upgrades.
- Conducting the lighting survey.
- Calculating lighting impacts.
- Calculating energy cost savings.
- Calculating maintenance cost savings.
- Performing the financial analysis.
- Measuring performance in trial installations.
- Evaluating occupant acceptance.

Although each of these steps can be performed using in-house staff, consider the services offered by various lighting professionals. These individuals may be independent consultants or employees of lighting management companies that provide turnkey services, including the survey, upgrade analysis, trial installation, full-scale installation, financing and ongoing maintenance. Although product vendors can assist with the assessment tasks, their objectivity may be limited by the range of products they sell. Local electric utilities may offer a choice of services to assist their customers in implementing energy-efficiency and load management programs. Table 18-1 illustrates the apparent trade-off between the cost of acquiring lighting expertise and the resulting objectivity in product selection and technical approach.

Table 18-1. Acquiring lighting expertise. *Source: EPA Green Lights.*

Source	Typical Cost	Objectivity	Turnkey Services
In-house	Depends on salary/ benefits and amount of training needed	High	Survey, specification, possible installation and/or maintenance
Independent Consultant	$0.01-$0.10/sq.ft.*	High	Survey, specification, project management
Lighting Management Company	$0-$0.03/sq.ft.	Medium	Survey, specification, product sales, installation and maintenance
Product Vendors	$0	Low	Survey, product sales, field assistance
Utility Services	$0	High	Depends on incentive program

*Depends on scope of services offered, diversity and travel costs.

PRIORITIZING FACILITY UPGRADES

The first step in prospecting for lighting upgrade opportunities is to determine which facilities should be addressed first. Those facilities whose upgrades are expected to yield the highest profits are usually the first to be surveyed and upgraded. This prioritization may be the first step in planning a financing strategy, because the savings realized from these initial projects can be used to finance subsequent upgrade projects.

Other factors may be used in prioritizing lighting upgrade projects. For example, a highly visible corporate headquarters building may not yield the highest returns, but it may be the best place to start for gaining recognition as a good "corporate citizen."

There are a variety of factors that should be reviewed when prioritizing facility upgrades. These factors include ownership, square footage, age, annual electricity cost, utility rate, availability of utility rebates or incentives, and annual occupancy hours. Facility upgrades may be prioritized based on a combination of these factors.

CONDUCTING THE LIGHTING SURVEY

To begin the process of assessing a building's lighting upgrade opportunities, information must be collected regarding the existing lighting system. This step is extremely important, because all future decisions in the lighting upgrade assessment will be based on the initial data collected. Therefore, accuracy and completeness are essential. As the existing lighting system data are collected, ideas for lighting upgrades may also be conceived.

The task of data collection can be broken down into three distinct steps.

1. Collect "project-wide" information.
2. Define prototype spaces.
3. Determine project quantities.

Collecting "Project-wide" Data

The first step in conducting the lighting survey is to collect "project-wide" data through a series of interviews with the building manager, electrician, utility specialist and accounting contacts. The information to be collected in the interviews consists of the agreed-upon lighting upgrade preferences, facility information and the assumptions to be used later in the financial analysis step.

Lighting Upgrade Preferences
The decision-maker in the organization may have preferences regarding the lighting system that will affect the final lighting upgrade specification. The following questions need to be answered before more

detailed information is gathered.

* What illuminance (light level) targets should be used?

* Is task/ambient lighting acceptable?

* What levels of visual comfort, color rendering and color temperature are desired?

* Is retrofitting preferred over fixture replacement?

* Are there specific technology preferences?

* Is group relamping and cleaning a viable maintenance option?

Clearly, this step in the process is an opportunity for the surveyor to suggest improved lighting technologies and practices. In some cases, the benefits of energy-efficient high quality lighting may have to be sold even before the survey begins!

Facility Information
Other project-wide data must be collected that addresses the operation and physical characteristics of the facility. The types of data to collect are:

* Approximate lighting system operating hours.
* Floor plans and square footage.
* Future plans for the building's use.
* Age of building and upgrade history.

The facility information will assist in the subsequent lighting system survey and will influence the selection of upgrade technologies. The age of the building is important to know because if the building was built before 1980, there is a good chance that some of the ballasts contain PCBs, which may require special handling at additional cost. (Chapter 20 provides guidelines for disposing of PCB-containing ballasts and mercury-containing lamps.)

Financial Analysis Assumptions
To present a credible financial analysis to the decision-maker, the following input assumptions must be agreed upon:

- Electricity rates and structure (including energy and demand charges).

- Labor rates for installation and maintenance.

- Inflation rates for materials and labor.

- Marginal income tax rates.

- Availability and values for rebates.

- Financial analysis terms (years of cash flow, discount rate).

- Costs for lamp and ballast disposal.

So far, the data collection step has consisted mainly of asking questions and recording answers. These answers will be used later in the analysis step.

Defining Prototype Spaces for Unique Upgrade Solutions
The second step in conducting the lighting survey is to perform a facility "walk-through," identifying the various locations where unique lighting upgrade solutions would be applied. In highly homogenous, modern office buildings, relatively few upgrade definitions would be needed. However, in older buildings with a variety of space types (such as hospitals or schools), a much larger number of lighting upgrade definitions will need to be determined. To define these prototype spaces, follow the steps outlined below.

1. Identify Unique Fixture Types
Most facility walk-throughs begin by looking up, observing the variety of fixture types in use. Before taking an inventory of the fixtures, begin by assigning a fixture name to each unique fixture configuration as defined by the characteristics listed below. This data can be organized in a fixture schedule as shown in Table 18-2.

- *Type and size of fixture* (e.g., 2x4 lens, 1x4 deep-cell parabolic, etc.).

- *Number of lamps per fixture.*

- *Specular reflector installed?*

- *Air handling fixture?* (Is the fixture used as part of the air distribution system?).

- *Type of lamp* (e.g., F40T12/ES, F96T12/741, 250W mercury vapor, etc.).

- *Type of ballast* (e.g., energy-efficient magnetic, hybrid, etc.).

- *Number of lamps per ballast.*

- *Fixture condition, comments* (condition of lens/louver/reflector, indicating those components that have deteriorated and should be upgraded/replaced).

To accurately determine the lamp and ballast types, it may be necessary to physically inspect a sample of luminaires. Alternatively, maintenance records may indicate the types of lamps and ballasts in use. If a mix of energy-saver and standard technologies are used, apply a weighting factor to the wattage and lumen values that represents the relative proportions of each technology.

2. *Estimate The Hours of Lighting Use*
Within each fixture type, a further distinction would be based on the hours of lighting use. For example, the lighting upgrade solution in a space with longer hours of operation may also include occupancy sensors.

3. *Indicate Whether Daylight Is Available*
Spaces with natural lighting may be candidates for daylight dimming or switching controls. As a result, these spaces may need to be analyzed separately, considering the various ballast and control options.

4. *Identify the Task-specific Light Level Targets*
Because the light-level targets ultimately drive the lighting upgrade solution, spaces with different visual tasks or illumination requirements should be analyzed separately.

5. *Indicate Whether Partitions Are Used*
Further defining the prototype spaces for lighting upgrades, the use of partitions may dictate a unique solution involving undershelf task lights or workspace-specific lighting systems.

6. *Indicate Areas with Different Physical Features*
Usually, the unique spaces for prototype upgrade solutions are

Table 18-2. Typical fixture schedule format. *Source: EPA Green Lights.*

Fixture ID	Fixture Type	Lamps/ Fixture	Reflector (Y/N)	Air (Y/N)	Lamp Type	Lamp Watts	Ballast Type	Lamps/ Ballast	Comments
A	2x4-L	4	N	Y	T12/CW	34	EE-Mag	2	good condition
B	1x4-P32	1	N	N	T12/CW	40	EE-Mag	1	good condition
C	2x2-_	2	N	Y	T12/CW U-lamp	40	EE-Mag	2	good condition

defined with the previous five characteristics. In other words, it is unusual for a space with the same fixture type, the same operating hours, the same daylight control options, the same light level targets and the same use of partitions to have different physical features. Nonetheless, there are cases in which these spaces are further distinguished by a number of unique physical features, including the ceiling height, task height and room colors. These factors could influence a slight change in the upgrade design.

The example in Table 18-3 illustrates how the above procedure is used to define prototype spaces for unique lighting upgrade solutions. In this example, there are six prototype spaces in which unique lighting upgrade solutions would be analyzed. Note that the last column in this table indicates a prototype identifier that is linked to the fixture

Table 18-3. Example of prototype space definitions.

Step 1	Step 2	Step 3	Step 4	Step 5	Step 6	Finish
Fixture ID	Hr/Yr	Daylight?	FC Target	Partitions?	Location/ Comments	Prototype ID
'A' (2x4-3L)	2500	Y	50	N	pvt office	A1
"	3500	N	50	N	pvt office	A2
"	"	"	"	Y	open office	A3
"	5000	N	15	N	hallway	A4
'B' (1x4-1L)	4500	N	15	N	rest room	B
'C' (2x2-2L)	2500	N	40	N	conf room	C

type. The example shows that the dominant fixture in the building (2x4 fixture) is used in four of the defined prototype spaces, meaning that up to four unique lighting upgrade analyses would be performed for upgrading spaces currently using this fixture. Note, however, if the presence of windows or partitions do not affect the lighting upgrade specification, fewer prototype spaces need to be defined and separately analyzed.

Determining Project Quantities

After the walk-through is complete and the analysis of prototype spaces has been performed, it is time to define the quantities of each upgrade component to determine the project cost and savings values. This third and final step in the lighting survey may involve a thorough room-by-room survey, or equipment quantities may be derived from existing as-built drawings (with some field verification).

Fixture Counts by Prototype Space

Using the fixture IDs defined in the above step, determine how many of each type of fixture exists in each prototype space. When performing a room-by-room survey, first identify the room with a unique name or identifier. Then identify the prototype space and indicate the number of fixtures in the room. If the prototype space identifier is linked to the fixture type (e.g., prototype space A3 uses fixture type A), then no more information needs to be collected regarding the fixtures.

Identify Needs for Additional Equipment

While visiting each space, determine whether task lights and/or occupancy sensors should be added. A visual inspection of each space type is needed in order to determine the correct mounting and type of task lights and occupancy sensors. Refer to the application guidelines for task lights (in Chapter 4) and for occupancy sensors (in Chapter 11). In our example, the room survey form might appear as shown in Table 18-4.

Optional Survey Data

In some cases, additional data may need to be collected to calculate existing and proposed footcandles. These data would include room dimensions, room surface reflectance and luminaire mounting height above the task.

Table 18-4. Typical survey form for defining project quantities.

Room ID	Prototype ID (Fixture/Space)	# Fixtures	#/Type of Occupancy Sensors Needed*	#/Type of Task Lights Needed*
101	A3	24	4/ceiling-IR	20 CFL undershelf
102	A1	4	1/switch-US-manual	2 CFL desklamps

Use Light Meters with Caution!

It is tempting to bring a light meter along during the survey to conduct spot checks of illuminance on task surfaces. Although it is useful to get an idea of the current illuminance values, bear in mind that it may be difficult to translate the actual reading into an average maintained illuminance value. For example, if the system had just been cleaned and relamped, the measured footcandle values would be much higher than the average maintained values. Conversely, if the lamps are living their last hour and the luminaires have never been cleaned, the light meter will show a value that is lower than the average maintained illuminance value. Because it is difficult to know where on the depreciation curve the system is currently operating, it is usually more accurate to calculate footcandles than to measure them as the survey is conducted. Alternatively, "initial" light levels can be measured if the luminaires are cleaned and new lamps are burned in for 100 hours; these initial readings can then be multiplied by the light loss factors (lamp lumen depreciation and luminaire dirt depreciation) to arrive at the true maintained light level in the space.

CALCULATING LIGHTING IMPACTS

The first important step in lighting upgrade analysis is to determine the maintained light levels needed for performing the visual tasks in the space. The lighting upgrades should be chosen based on their ability to most efficiently deliver the target light level. Therefore, it is essential that each lighting upgrade analysis includes an evaluation of the impacts on light levels. There are two types of light level calculations: *Relative Light Output* and *Illuminance Calculations*. Most lighting analysis software will perform these calculations based on user-entered application data.

Relative Light Output (RLO)

This method is relatively simple and can be performed on virtually any lighting system. It compares the maintained lumen output of the baseline luminaires with the maintained lumen output of the upgraded system. The RLO is the percentage of the baseline maintained lumen output that will be provided by the new system. In other words, a 90 percent RLO means that about a 10 percent reduction in maintained illumination can be expected. This approach will produce reliable results *if the following conditions prevail:*

- *The type of light distribution will not be changed.* For example, the RLO calculation is valid if downlights are replaced with other downlights. Conversely, the RLO calculation will be misleading in applications where uplighting luminaires replace downlighting troffers.

- *The number and location of luminaires will not be changed.* The foot-candles delivered by the luminaires is a function of the room geometry and room surface reflectances. If these factors change, then the RLO result becomes invalid.

The RLO calculation requires a minimum amount of input data. Use the equations below to calculate the relative light output of lighting upgrades that meet the above criteria.

$$\text{Relative Light Output}\left(\text{RLO\%}\right) = \frac{\text{Upgrade Light Outlut} \times 100\%}{\text{Baseline Light Output}}$$

Where ...

Baseline Light Output	=	# lamps × initial lumens/lamp × ballast factor × luminaire efficiency × LLF*
Upgrade Light Output	=	# lamps × initial lumens/lamp × ballast factor × luminaire efficiency × LLF*

*LLF (light loss factor) = LLD (lamp lumen depreciation) × LDD (luminaire dirt depreciation)

Illuminance Calculation

Although the RLO method provides an approximation of the *relative* impacts of the lighting upgrade, it tells us nothing about the *absolute* impacts. Use the illuminance calculation to predict the average maintained footcandles in the space, and compare the results with the selected target illumination level.

To perform the illuminance calculation, photometric data is needed that describes the performance of the particular luminaires in the ceiling. Follow the example below in calculating the average maintained horizontal illuminance in the sample space. To begin, here is the formula that is used:

$$fc = \frac{\#lamps \times lumens/lamp \times CU \times BF \times LLF}{SF}$$

Where...

CU = coefficient of utilization
BF = ballast factor
LLF = lamp lumen depreciation × luminaire dirt depreciation
SF = room area in sq.ft.

The coefficient of utilization (CU) is the percentage of bare lamp lumens that strike the workplane. This value is looked up on a table provided on the luminaire's photometric report. In Table 18-5, note that a CU value is assigned for selected values of *room cavity ratio* and ceiling, wall and floor *reflectances*. Because the room geometry influences the CU, a value known as the *room cavity ratio* is calculated which describes the effect of the room geometry on the CU:

$$Room\,Cavity\,Ratio = \frac{5 \times h \times (L + W)}{(L \times W)}$$

Where ...

h = distance between the fixture and work surface
L = length of room in feet
W = width of room in feet

Table 18-5. Sample luminaire coefficient of utilization values.

| Floor Reflectance | 20% | | | | | | | | | | | | | |
| Ceiling Reflectance | 80% | | | | 70% | | | | 50% | | | 30% | | |
Wall Reflectance	70%	50%	30%	10%	70%	50%	30%	10%	50%	30%	10%	50%	30%	10%
RCR=1	76	73	71	67	74	72	70	68	69	67	66	66	65	64
RCR=2	71	67	63	60	69	65	62	59	63	59	58	61	59	57
RCR=3	66	60	56	52	64	59	55	52	57	54	51	56	52	50
RCR=4	61	54	49	46	60	53	49	45	52	49	45	50	47	44
RCR=5	56	49	43	40	55	48	43	39	47	42	39	46	42	39
RCR=6	52	44	39	35	51	43	39	35	42	38	34	41	37	34
RCR=7	48	40	35	31	47	39	35	31	38	34	30	37	33	30
RCR=8	44	36	30	27	43	35	30	27	34	30	26	34	29	26
RCR=9	41	32	27	23	40	32	27	23	31	26	23	30	26	23
RCR=10	38	29	24	21	37	29	24	21	26	23	20	28	23	20

Room reflectance values can be estimated. Light-colored ceilings generally have a reflectance value of 70-80 percent, while light-colored walls have a lower reflectance—usually about 40-50 percent. The most common floor reflectance is a "medium" value, which translates to an average reflectance of about 20 percent. Luminance meters can be used for accurately measuring reflectances.

CALCULATING ENERGY COST SAVINGS

Once the existing and proposed upgrade components have been defined for the prototype space, the next step is to calculate the electricity cost savings. Although this step is almost universally performed with software programs, this section describes the mechanics of the calculations. The calculation of electricity cost savings is performed in three steps:

1. Calculate peak electricity demand savings in kilowatts (kW).
2. Calculate energy savings in annual kilowatt-hours (kWh).
3. Calculate electricity cost savings in dollars per year ($/year).

To begin the calculations for a given lighting upgrade, determine the following values for both the existing and proposed lighting systems:

*   *Electrical Load (kW or Watts).* Refer to the system performance tables included in manufacturer literature. Convert watts to kilowatts (kW) by dividing watts by 1,000. Then multiply the fixture kilowatts by the number of fixtures in the defined spaces to determine the lighting system kilowatts.

*   *Estimated Hours of Use Per Year (Hours/Year).* The lighting system operating hours can be estimated or measured directly. Measurement options are discussed later in this chapter.

Step One: Calculate Peak Demand Savings
The savings in peak electrical demand that can be attributed to the lighting upgrade is calculated:

Peak Demand Savings (kW) = Baseline Peak Demand - Upgrade Peak Demand

How To Calculate Footcandles Using The Lumen Method

The following example illustrates the procedure for calculating the light level produced by a fluorescent troffer system (recessed in grid ceiling). Where suspended luminaires are used, an additional step is required for calculating the ceiling cavity ratio and the equivalent reflectance of the luminaire plane. Refer to IESNA literature for the complete procedure to calculate illuminance provided by suspended luminaires.

Step One: Gather Data - The data needed for calculating footcandles in our example is provided below:

Room: L=30 ft., W=12 ft.   Luminaires: 3-lamp deep-cell parabolic
Ceiling Height: 9 ft.   Luminaire Photometrics: See Table 18-5
Task Height: 2-1/2 ft.   Lamps: F32T8/735
Ceiling Reflectance: 70%   Initial Lamp Lumens: 2,850
Wall Reflectance: 50%   Lamp Lumen Depreciation (LLD): 0.91
Floor Reflectance: 20%   Luminaire Dirt Depreciation (LDD): 0.85
Luminaire Quantity: 6   Ballast Factor (BF): 0.88

Step Two: Calculate Room Cavity Ratio - The values above are used to calculate the room cavity ratio (RCR). This value indicates the relative shape of the room; RCR close to one indicates a relatively low ceiling compared to the room area, while a high RCR represents a tall ceiling compared to the room area. Rooms with high RCRs absorb a higher percentage of the lumens as they travel from the luminaire to the task surface.

RCR = 5 × h × (L + W) ÷ (L × W), where h=6.5, L=30, W=12
RCR = 5 × 6.5 × (30 + 12) ÷ (30 × 12) = 3.8

Step Three: Look Up Coefficient of Utilization - In our example, we assume that Table 18-5 provides the CU data for the luminaires in the sample space. Note that CU data can vary significantly between fixture types and among manufacturers; use the CU data that corresponds to the specific fixture in question. For approximate photometric data for common luminaire types, refer to the data provided in the IESNA *Lighting Handbook*. The CU for our particular example is found by interpolating the values in Table 18-5 for the room's surface reflectances. Because the RCR is between 3 and 4, look up both values; these are 59 (RCR=3) and 53 (RCR=4). The interpolated value for the 3.8 RCR value is approximately 59 – (0.8 × (59–

---

53)) = 54.2 or 54; expressed as a decimal, the CU is 0.54.

Step Four: Calculate Illuminance - We now have all of the values for calculating footcandles to be maintained by the defined lighting system.

fc = # lamps/fixture × # fixtures × initial lamp lumens × CU × BF × LLD × LDD ÷ (L × W)

fc = 3 lamps/fixture × 6 fixtures × 2,850 lumens × 0.54 × 0.88 × 0.91 × 0.85 ÷ (30 × 12)

fc = 52 fc, maintained

Baseline Peak Demand: electrical load (kW) × % of lighting on at building's peak demand × % dimming factor

Upgrade Peak Demand: electrical load (kW) × % of lighting on at building's peak demand × % dimming factor

It is not realistic to assume that 100 percent of the new lighting system will be energized at the point in time when the building's electrical demand reaches its highest point during the month. Therefore, percentage values should be applied to the lighting system's total kilowatt load for both the baseline and upgrade cases (as shown in the above equations) that provide an estimate of the percentage of the lighting system's connected wattage that is energized at the time when the building reaches its peak rate of electricity consumption. In addition, if dimming controls or light-level switching systems exist or are proposed, apply a *"peak dimming factor"* percentage value to compensate for the weighted average affects of dimming on peak demand.

Step Two: Calculate Annual Energy Savings
The annual energy savings in kWh is calculated:

Annual Energy Savings (kWh) = Baseline Energy Use - Upgrade Energy Use

Baseline Energy Use (kWh): electrical load (kW) × dimming factor (%) × hours/year

Upgrade Energy Use (kWh): electrical load (kW) × dimming factor (%) × hours/year

Although no demand coincidence factors are used in these equations, the energy use may need to be adjusted if dimming systems exist or are proposed. The dimming factor should represent the weighted average percent of full-load power that the lighting systems consume on an annual basis.

Step Three: Calculate Annual Energy Cost Savings

The components of energy cost savings attributable to a lighting upgrade are savings in peak demand charges (in $/kW/month) and savings in energy use charges (in $/kWh). Although most major commercial and industrial electricity rates include charges for both demand and energy use, smaller accounts may not be subject to peak demand charges. To calculate the energy cost savings, use the following equations as applicable.

Annual Electricity Cost Savings ($/yr) = Peak Demand Cost Savings +
Energy Cost Savings

Peak Demand Cost Savings = Peak Demand Savings (kW) × demand rate
($/kW/mo) × 12 mo/yr

Energy Cost Savings = Energy Savings (kWh) × electricity rate ($/kWh)

Further complicating this calculation are the impacts of complex rate structures that charge different rates for peak demand and/or energy use based on the *time of day*. To accurately calculate the electricity cost savings with "time-of-use" rates, perform separate annualized calculations as described above for each daily time period defined in the rate schedule. These individual costs should then be aggregated to the total annual energy cost savings. If seasonal rates are used, calculate the annual weighted average electricity rate based on the number of months that each seasonal rate is in effect.

Finally, unique "ratcheted" demand charges can delay a portion of the anticipated cost savings. With this unique rate structure, the demand charge is calculated based on a percentage (up to 100 percent) of the facility's peak demand registered during the past 12 months. Under this type of rate, the demand charge savings may not begin to be realized until the following year. For assistance in calculating electricity cost savings with these types of rate structures, contact the local utility.

Impacts on HVAC Use

Because lighting systems emit heat into the conditioned spaces of buildings, they contribute to the amount of heat that must be removed by the air conditioning system. In the northern latitudes, the heat from the lighting system may reduce the amount of energy needed to heat the building. However, for most buildings in the southern latitudes, the building warm-up period is completed during the early morning before the lights have been turned on and, therefore, the lights do not contribute much to help the building warm up.

In most buildings, energy-saving lighting upgrades will yield savings in air conditioning costs, because the new system produces less heat to be removed by the air conditioning equipment. Although there may be a heating "penalty" resulting from lighting upgrades in many buildings, this penalty is usually small in comparison to the air conditioning savings. However, in northern latitudes, the heating penalty could equal the air conditioning savings, canceling the HVAC interaction effect on overall lighting upgrade energy savings.

Most lighting upgrades will result in HVAC energy (and demand) savings adding 0-30 percent to the savings attributable to the lighting energy savings. In new construction applications, a low-wattage lighting design can yield huge savings in the installed costs of the cooling and air distribution systems.

Figure 18-1. All of the energy that is consumed to produce light is eventually converted to heat. Therefore, reductions in lighting energy use will produce reductions in air conditioning costs. These savings can be much higher than any heating "penalties" resulting from a lighting upgrade, particularly in the southern latitudes. *Courtesy: National Lighting Bureau.*

Calculation Form

Use the form in Figure 18-2 (3 pages total) to calculate the approximate energy cost savings expected from lighting upgrades.

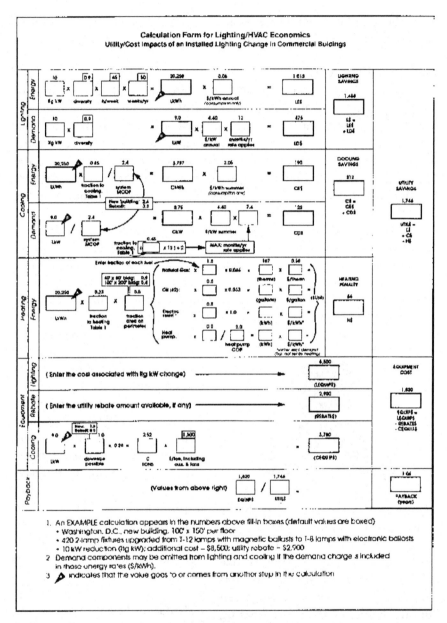

Figure 18-2. Calculation form, definitions and data table for Lighting/ HVAC Economics. *Courtesy: American Society of Heating, Refrigerating & Air-Conditioning Engineers, Inc.*

List of Inputs in Order of First Occurrence

• **Ltg kW** The *reduction* in lighting input kW due to a proposed retrofit or design change.

• **Diversity** The average fraction of lights turned on during the period that lights are normally on, accounting for random absences. (Default = 0.9)

• **Hr/week and weeks/yr** Hours per week and weeks per year that lights are normally on. (Defaults = 45 hr/week and 50 weeks/yr)

• **$/kWh, annual** The average cost per kWh, including consumption but not demand charges. Use marginal rate, (i.e., the savings due to any decreased kWh).

• **Months/yr rate applies** For lighting demand, the number of months the utility charges the demand rate entered in the next input.

• **$/kW, annual** The cost per kW electric demand applying for Months/yr in the previous input.

• **Fraction to cooling and fraction to heating, from Table 1** Look up values for the city in *Table 1* with the climate most similar to the city being studied.

• **System MCOP** The seasonal marginal coefficient of performance of cooling equipment, including all auxiliaries and supply and return air fans. This input indicates not only how efficient equipment is, but also how much of equipment can "participate" in any load reduction—higher MCOP indicates less equipment participating, as in a retrofit. Note that small and large buildings have similar MCOPs, because higher plant efficiencies in large buildings are offset by higher distribution energy. (Default = new building: 2.4, retrofit: 3.0)

• **$/kW, summer** The cost per kW electric demand applying for cooling season months.

• **Fraction area on perimeter** The fraction of floor area that is within 15 ft of a perimeter wall. It is assumed that only this area has heat load that may be offset by lighting heat, so larger buildings have a smaller heating impact of a lighting change. (Examples—40' x 60' floor: 0.9; 100' x 200' floor: 0.4)

• **COP, heat pump** Seasonal heat pump COP. (Default = 2.0)

• **LEQUIP$** The cost of lighting equipment associated with the installed change Ltg kW.

• **REBATE$** The utility rebate amount available to offset the cost of lighting equipment.

• **Downsize possible** Fraction of theoretical cooling equipment downsizing (due to lighting reduction) that can be implemented in the job at hand. One may use one of the extremes shown for the default, or a value in between if some portion of the reduced lighting will be considered in cooling sizing. (Default = new building: 1.0; retrofit: 0.0)

• **$/ton, incl. aux. & fans** The cost of cooling equipment including all associated pumps, cooling towers, supply and return air fans, etc. (Default = $1,500/ton)

Derivation of Constants Appearing in the Calculation, in Order of Appearance:

• **12 and 2** (in Cooling Demand): 12 = months/year; 2 = months beyond the normal cooling season that cooling is assumed to set the peak electric demand for a month.

• **0.045** (in Heating Energy): (3,413 Btu/KWh)/(100,000 Btu/therm x 0.75 seasonal efficiency) = 0.045 therms (fuel required)/kWh (lighting removed).

• **0.033** (in Heating Energy): (3,413 Btu/kWh)/(137,000 Btu/gallon x 0.75 seasonal efficiency) = 0.033 gallons (fuel required)/kWh (lighting removed).

• **0.28** (in Equipment, Cooling): (3,413 Btu/hr/kW)/(12,000 Btu/hr/ton) = 0.28 tons (cooling load)/kW (lighting removed).

Figure 18-2 (continued). *Courtesy: American Society of Heating, Refrigerating & Air-Conditioning Engineers, Inc.*

Table 1. Fraction of Annual Lighting Heat to Cooling and Heating

Location	Fraction Lighting to: Cool	Heat	Location	Fraction Lighting to: Cool	Heat	Location	Fraction Lighting to: Cool	Heat
ALABAMA			**LOUISIANA**			**OHIO**		
Birmingham	0.57	0.06	Alexandria	0.64	0.03	Cincinnati	0.42	0.25
Mobile	0.52	0.01	Lake Charles	0.68	0.02	Cleveland	0.38	0.31
Mobile	0.60	0.01	New Orleans	0.69	0.02	Columbus	0.41	0.27
Montgomery	0.61	0.04	Shreveport	0.61	0.05	Dayton	0.42	0.29
ARIZONA			**MAINE**			Toledo	0.37	0.33
Flagstaff	0.32	0.37	Portland	0.27	0.36	**OKLAHOMA**		
Phoenix	0.71	0.00	**MASSACHUSETTS**			Altus	0.54	0.02
Tucson	0.69	0.02	Boston	0.34	0.27	Enid	0.49	0.16
ARKANSAS			Springfield	0.35	0.35	Oklahoma City	0.51	0.17
Blytheville	0.51	0.16	**MICHIGAN**			Tulsa	0.51	0.17
Fort Smith	0.53	0.14	Detroit	0.33	0.31	**OREGON**		
Little Rock	0.54	0.11	Grand Rapids	0.33	0.35	Burns	0.30	0.35
CALIFORNIA			Lansing	0.34	0.32	Eugene	0.26	0.14
Barstow	0.56	0.10	Sault Sainte Marie	0.22	0.41	Medford	0.37	0.19
Bishop	0.53	0.15	Traverse City	0.29	0.38	Pendleton	0.36	0.24
Los Angeles	0.56	0.00	**MINNESOTA**			Portland	0.27	0.14
Sacramento	0.49	0.04	Duluth	0.22	0.42	**PENNSYLVANIA**		
San Diego	0.52	0.00	International Falls	0.24	0.44	Philadelphia	0.41	0.24
San Francisco	0.38	0.02	Minneapolis	0.33	0.39	Pittsburgh	0.38	0.30
Santa Barbara	0.21	0.05	**MISSISSIPPI**			Scranton	0.35	0.34
COLORADO			Biloxi	0.65	0.01	Williamsport	0.36	0.31
Colorado Springs	0.37	0.29	Columbus	0.59	0.10	**RHODE ISLAND**		
Denver	0.39	0.29	Jackson	0.61	0.07	Providence	0.32	0.29
Grand Junction	0.41	0.29	**MISSOURI**			**SOUTH CAROLINA**		
Trinidad	0.44	0.30	Columbia	0.44	0.26	Charleston	0.62	0.05
DELAWARE			Kansas City	0.44	0.22	Columbia	0.58	0.09
Dover	0.41	0.23	St. Louis	0.46	0.25	Myrtle Beach	0.56	0.08
Wilmington	0.41	0.32	Springfield	0.47	0.22	**SOUTH DAKOTA**		
FLORIDA			**MONTANA**			Huron	0.35	0.41
Miami	0.87	0.00	Billings	0.32	0.36	Rapid City	0.34	0.36
Jacksonville	0.72	0.02	Glasgow	0.30	0.41	Sioux Falls	0.35	0.39
Orlando	0.80	0.00	Great Falls	0.29	0.37	**TENNESSEE**		
Pensacola	0.66	0.01	Helena	0.27	0.38	Knoxville	0.50	0.16
Tampa	0.80	0.00	**NEBRASKA**			Memphis	0.53	0.13
GEORGIA			Grand Island	0.39	0.34	Nashville	0.49	0.14
Atlanta	0.57	0.10	North Platte	0.38	0.35	**TEXAS**		
Augusta	0.61	0.08	Omaha	0.40	0.32	Amarillo	0.49	0.19
Macon	0.60	0.06	**NEVADA**			Corpus Christi	0.74	0.01
Savannah	0.66	0.05	Ely	0.35	0.36	Dallas	0.60	0.07
Valdosta	0.67	0.02	Las Vegas	0.61	0.05	Houston	0.73	0.02
IDAHO			Reno	0.36	0.25	Lubbock	0.53	0.14
Boise	0.34	0.26	Winnemucca	0.39	0.30	San Antonio	0.71	0.03
Lewiston	0.33	0.24	**NEW HAMPSHIRE**			**UTAH**		
Pocatello	0.30	0.35	Manchester	0.33	0.37	Salt Lake City	0.34	0.29
ILLINOIS			**NEW JERSEY**			Wendover	0.37	0.29
Champaign	0.41	0.31	Trenton	0.40	0.28	**VERMONT**		
Chicago	0.36	0.33	**NEW MEXICO**			Burlington	0.29	0.39
Peoria	0.42	0.31	Alamogordo	0.56	0.14	**VIRGINIA**		
Rockford	0.36	0.35	Albuquerque	0.47	0.20	Richmond	0.46	0.16
INDIANA			Clovis	0.51	0.17	Roanoke	0.46	0.18
Fort Wayne	0.39	0.32	**NEW YORK**			**WASHINGTON**		
Indianapolis	0.41	0.29	Albany	0.34	0.35	Seattle	0.16	0.17
South Bend	0.37	0.32	Buffalo	0.33	0.34	Spokane	0.27	0.32
Terre Haute	0.43	0.32	Syracuse	0.34	0.35	**WEST VIRGINIA**		
IOWA			New York City	0.35	0.24	Charleston	0.45	0.22
Council Bluffs	0.40	0.32	**NORTH CAROLINA**			Clarksburg	0.39	0.30
Mason City	0.34	0.39	Greensboro	0.49	0.16	**WISCONSIN**		
Sioux City	0.38	0.36	Raleigh	0.52	0.14	Green Bay	0.30	0.39
KANSAS			Wilmington	0.58	0.05	Madison	0.35	0.38
Dodge City	0.44	0.26	**NORTH DAKOTA**			Milwaukee	0.36	0.36
Goodland	0.41	0.30	Bismarck	0.32	0.42	**WYOMING**		
Kansas City	0.44	0.22	Fargo	0.29	0.42	Casper	0.33	0.37
Wichita	0.47	0.19	Grand Forks	0.29	0.42	Cheyenne	0.32	0.35
KENTUCKY			Minot	0.28	0.42	Rock Springs	0.29	0.40
Covington	0.42	0.25				**WASHINGTON, DC**	0.45	0.23
Hopkinsville	0.49	0.17						
Louisville	0.48	0.22						

Note: Heating fraction is of perimeter zone lighting only
Note: Table cooling values assume economizer operation; if no economizer, see below:

Size Building	Use for "Frac...to Cool"	Example (Washington, DC)
Small (2,000 ft² per floor)	0.2 + 0.8 × "Frac...to Cool" in Table	0.2 + 0.8 × 0.45 = 0.56
Large (20,000 ft² per floor)	0.5 + 0.5 × "Frac...to Cool" in Table	0.5 + 0.5 × 0.45 = 0.73
Intermediate size	Weighted average of above	(0.56 + 0.73)/2 = 0.65

Figure 18-2 (continued). *Courtesy: American Society of Heating, Refrigerating & Air-Conditioning Engineers, Inc.*

This form incorporates calculations to determine savings in demand charges and energy charges resulting from reduced lighting power requirements and their impacts on HVAC use. In addition, this form allows users to determine reductions in cooling equipment costs for new construction when lighting efficiency measures are included in the design. For more information about this calculation, contact the Electric Power Research Institute (EPRI) or the American Society of Heating, Refrigerating & Air-Conditioning Engineers (ASHRAE).

Clearly, projections of energy cost savings are based on a number of estimates. Some of these estimates have a greater impact on the final calculated value. Table 18-6 lists these assumptions in descending order of their typical impact on bottom-line calculations. Note that the impacts of these factors can vary depending on the technologies considered and local utility rate structures.

Table 18-6. "Estimated" factors that affect energy cost calculations.

Factors Used In Energy Cost Calculations	Methods To Determine Factor (Or Improve Accuracy)
Annual operating hours (hr/yr)	Measure lighting circuit operating hours with logging devices.
Coincidence of peak demand (%)	Measure kW versus time for lighting and HVAC systems; use lighting kW value at building peak demand and divide by connected lighting kW.
Dimming factor (%)	Make assumption based on weighted average of occupant preferences for dimming level; refer to manufacturer-supplied dimming efficacy curves.
Weighted average time-of-use electricity rates ($/kWh)	Measure distribution of lighting hours among rate periods with a time-of-use lighting logger; calculate weighted average rate based on this distribution and the time-of-use rates.
Thermally corrected (W) wattage values	Use ANSI wattage correction factors listed in Appendix II.

CALCULATING MAINTENANCE COST SAVINGS

Savings in average annual lighting maintenance costs can be realized as a by-product of the lighting upgrade. The following are the most common sources of maintenance cost savings.

• Reduced lamp costs due to bulk purchasing for group relamping.

• Reduced ballast material costs due to tandem wiring.

• Reduced material costs due to longer equipment life (with controls reducing operating hours).

• Reduced relamping and cleaning costs due to initiation of group maintenance program.

Software programs calculate these impacts on lighting system maintenance costs. It is possible, however, that with the use of the more expensive triphosphor lamps and the use of instant-start ballasts, lamp material costs *could* increase. However, such increases are usually quite small. To achieve the biggest impact on maintenance savings, initiate a program of planned maintenance whereby luminaires are cleaned and relamped at regular intervals—at approximately 70 percent of rated lamp life (see Chapter 20).

PERFORMING THE FINANCIAL ANALYSIS

Financial profitability is one of the key goals of any lighting upgrade. For decades, energy conservation projects have been cost-justified on the basis of simple payback (initial cost divided by annual cost savings). Although this calculation portrays an accurate relationship between installed costs and the first year's cost savings, it is risky to use simple payback for making final investment decisions. Two other measures—*Internal Rate of Return* and *Net Present Value*—overcome these limitations and provide financial decision-makers with an objective assessment of the lighting upgrade project's value in comparison to other competing capital investments.

Because the simple payback calculation ignores the potentially significant changes in net savings that occur after the project has paid for itself, simple payback provides an unreliable measure of profitability. In addition, the simple payback method ignores the time-value of money—an important consideration when deciding on the best investment options for limited capital funds. This section describes how to develop a reliable assessment of a lighting upgrade project's profitability—with the help of a personal computer.

Step One: Estimate Cash Flows Over 10 To 20 Years
The key to a reliable financial analysis is to determine the net cash flow forecast over 10 to 20 years for each of the lighting upgrade options under consideration. The net cash flow is the year-by-year difference in owning and operating costs between the baseline system and the upgrade system. A typical 10-year (short-term) net cash flow projection is illustrated in Table 18-7. Note that the $59,760 initial cost and the other costs for energy and maintenance are shown as negative values; the net savings produced by the upgrade are shown as positive values.

The cash flow projection should take into account *all* cash flows *when* they occur, such as:

- Material cost.
- Installation cost.
- Electricity costs.
- Maintenance costs.
- Financing costs.

- Waste disposal costs.
- Consulting fees.
- Project management costs.
- Inflation factors.
- Tax effects.

As shown in Table 18-8, the selection of the analysis term (in years) will have a significant impact on the results of the life-cycle cost analysis. Choosing too short of a term will favor short-sighted low-cost/low-profit investments—known as "cream-skimming." Choosing too long of an analysis term is risky only if the business is likely to move or sell the property before the end of the analysis term. (Note, however, that lighting upgrades and efficiency improvements in general can increase the owner's net operating income in leased facilities, thereby significantly increasing the value of the building prior to selling it.)

Step Two: Determine The Hurdle Rate
Every organization must decide what degree of profitability is required to justify an investment. The "hurdle rate" is a percentage value that defines the minimum rate of return required for an investment to be defined as "profitable." An organization's investment hurdle rate may be

Table 18-7. Example cash flow table.

| Year | Existing Cash Flow | | Proposed Cash Flow | | Net Cash Flow |
	Energy Costs	Maintenance Costs	Energy Costs	Upgrade Cost & Maintenance Costs	
0				-59,760	-59,760
1	-21,750	-7,661	-7,679	-2,238	19,464
2	-22,402	-7,891	-7,909	-2,305	20,078
3	-23,074	-8,128	-8,147	-2,374	20,681
4	-23,766	-8,372	-8,391	-3,462	20,285
5	-24,479	-8,623	-8,643	-3,566	20,893
6	-25,214	-8,881	-8,902	-3,673	21,520
7	-25,970	-9,148	-9,169	-3,783	22,166
8	-26,749	-9,422	-9,444	-3,897	22,831
9	-27,552	-9,705	-9,728	-4,013	23,516
10	-28,378	-9,996	-10,019	-4,134	24,221

Table 18-8. Effect of analysis term on profitability.

Analysis Term (Years)	Net Present Value ($)	Internal Rate of Return (%)
1	-42,355	-67.4%
2	-26,349	-23.5%
3	-11,628	0.4%
4	1,263	13.0%
5	13,118	20.4%
6	24,021	25.1%
7	34,048	28.0%
8	43,269	30.0%
9	51,749	31.4%
10	59,548	32.4%
11	66,719	33.0%
12	73,315	33.5%
13	79,381	33.9%
14	84,959	34.1%
15	90,089	34.3%
16	94,806	34.5%
17	99,145	34.6%
18	103,135	34.7%
19	106,804	34.7%
20	110,179	34.8%

influenced by the current cost of capital (effective borrowing rate), opportunity costs and/or the risk of the investment. For low-risk investments in lighting, the hurdle rate is typically chosen between 8 percent and 20 percent.

**Step Three: Calculate Internal Rate of Return (IRR) and
Net Present Value (NPV)**

With a few keystrokes, a personal computer can quickly calculate the profitability of a project given its net cash flow projection. All spreadsheet programs and many lighting analysis programs are capable of calculating both net present value and internal rate of return. These profitability measures take into account the time-value of money and the cash flow that occurs over the period when the investor expects to receive the benefits of the investment. Table 18-9 lists the values needed to calculate NPV and IRR. The initial IRR "guess" provides a starting point for the computer's iterative computation of IRR.

Net Present Value (NPV)

The NPV of a project is expressed as a dollar amount, which represents the present value (or present "worth") of the cash flow projection. Using the hurdle rate as the "discount rate," future cash flows are discounted to determine their value to the investor *today*. The discount rate is essentially the opposite of an interest rate—an interest rate operating in reverse. *Interest rates* are used to calculate the *future* value of today's investment; *discount rates* are used to calculate the *present* worth of a future value. To calculate the NPV, the software program will ask for the discount rate; enter the chosen hurdle rate of 8-20 percent.

The NPV dollar amount represents the amount of money *in cash today* that is equivalent to the net value of the project's financial performance over time, *including the cost of the investment*. If the NPV is a positive dollar amount, the project is profitable (i.e., the IRR has cleared the hurdle rate). If the NPV is a negative dollar amount, then the project is, by definition, not profitable. Therefore, the NPV measure not only tells us whether the project is profitable, but *how* profitable it is. When prioritizing investments, the project with the highest NPV is the most profitable and should receive first priority, assuming there are no limitations in acquiring capital.

Table 18-9. Computer input values for calculating NPV and IRR.

Profitability Measure	*Computer Inputs Needed*
Net Prevent Value (NPV)	Year-by-year cash flow; discount rate
Internal Rate of Return (IRR)	Year-by-year cash flow; IRR guess

Internal Rate of Return (IRR)

The IRR is a single value, expressed as a percentage, that is compared with the organization's hurdle rate to determine if the project is profitable. Like the NPV calculation, the IRR is based on a project's net cash flow projection over time. However, the IRR does not require the input of a hurdle rate; it is simply a value that represents the rate of net financial return inherent in the cash flow.

The IRR is actually the discount rate at which a project's NPV equals zero. Therefore, the values of IRR and NPV are related. When the project's NPV equals zero, the IRR equals the organization's hurdle rate used in the NPV calculation. When the NPV is greater than zero, the IRR exceeds the hurdle rate. Note that the IRR does not indicate how profitable the project is; it only indicates whether or not the project is profitable.

MEASURING PERFORMANCE IN TRIAL INSTALLATIONS

Product literature and advertisements cite the amazing benefits of improved lighting performance and energy cost savings that are expected from lighting upgrade technologies. Although most claims in lighting product literature are true under specific conditions, they may not be true in yours. How can these claims be tested before investing great sums of money in a facility-wide upgrade? This section outlines procedures for measuring the performance of lighting upgrades in a trial installation.

Trial installations in an area of 5,000-15,000 sq.ft. can be specified, purchased and installed in a relatively short time frame. The upgrade's impacts on light level, energy consumption and hours of operation can be determined by taking measurements before and after the installation.

Evaluating Illumination Performance
One of the most critical variables to measure in a trial installation is the retrofit lighting product's performance in delivering illumination to the visual task. To accurately assess light levels *before and after* a trial upgrade, follow the steps outlined below:

Start with New Lamps and Clean Fixtures
Light output can be affected by the age of the lamps and dirtiness of the fixture. The baseline light level readings should be made only after the existing fixtures in the trial installation area have been cleaned and the existing lamps have been replaced with new lamps (same wattage

and type used in existing system). *To accurately measure "initial foot-candles", allow for a 100-hour lighting operation "burn-in" period before taking measurements as described below.*

Allow Time for System Warm-up
    Most lighting installations take some time to reach a stable condition after switch-on. Twenty minutes are typically sufficient for installations using fluorescent lamps, and it may take even longer for some HID lamps.

Eliminate Daylight Effects
    Daylight and sunlight can produce very large variations in lighting. To evaluate an electric lighting installation without any daylight contribution, take measurements after dark or with the blinds closed.

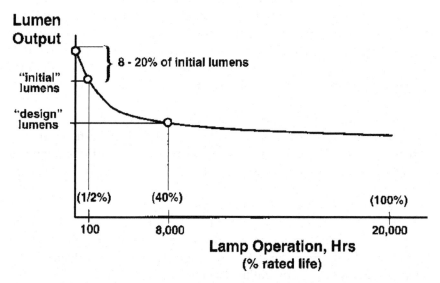

Figure 18-3. Do not measure light levels just after installing new lamps. Over the first 100 hours, significant lamp lumen depreciation occurs before the "initial" lumen output is reached. The lumen output of a lamp when it reaches 40 percent of its rated life represents the average lumen output that it will deliver over its life. *Courtesy: EPA Green Lights.*

Check Supply Voltage

The light output of most lamps is directly affected by the supply voltage. At the time of the survey, measure the supply voltage to verify that it is not below acceptable levels (check with the electric utility).

Properly Position the Light Meter

When making light level measurements, put the illuminance meter at the proper height on the work surface, and be careful not to shadow the meter by holding it close to your body. Also, be careful to avoid reflections off clothing that could influence the measurement.

Record Light Level Readings

Use the light meter to measure the footcandles at a variety of workplane locations including specific task locations and randomly throughout the room, between fixtures and in corners. Also measure illuminance on vertical surfaces, if applicable. Be certain to record the *locations* of readings for the baseline case so the procedure can be repeated when evaluating the upgrade in a trial installation. (Adhesive labels can be used to mark measurement locations and corresponding values.)

Calculate Average Maintained Light Level

The average light level measured in the room should be adjusted to account for anticipated lamp and dirt depreciation effects to determine the average *maintained* light level. The light loss factor to be applied to the initial readings is the product of the lamp

Figure 18-4. Illuminance meters can measure light levels in footcandles (fc) or lux (lx). This meter can be reconfigured to read either value. *Courtesy: Minolta.*

lumen depreciation (LLD) and the luminaire dirt depreciation (LDD) factors. (Refer to Chapter 20 for definitions and listings of typical light loss factors.) Multiply the initial light level readings by the light loss factor to determine the average maintained light level, and compare this value with the target light level.

Repeat Procedure

Repeat the above procedure after the trial installation is complete. The upgrade's light-level performance should be measured under the *same* conditions as in the base case.

Measuring Wattage

To verify that lighting upgrades will deliver the claimed wattage reductions, directly measure the wattage of lamp/ballast systems using a hand-held watt-meter. Note that this procedure involves exposure to line voltage (120V or 277V) and that a trained electrician should perform the measurement. The power to the circuit should be turned off while the connections are being made.

Most hand-held watt-meters require access to the ballast's power leads for the device to measure voltage, amperage and power factor, and to display the calculated wattage. To measure the wattage of a lamp/ballast system, first identify the hot lead (black wire) that is connected to the ballast. To measure the relatively small amount of current being used by the ballast, it will be necessary to place the current sensing transducer over several windings of the black wire. If enough slack wire exists, wind the wire into at least four loops and place the current transducer over the loops. Then remove the wire nuts that connect the ballast to the incoming electrical power supply and attach one voltage clip to each wire.

After all connections have been made, turn on the power and read the wattage on the display. (If the resulting wattage is negative, reverse the orientation of the current transducer.) Divide the displayed wattage by the number of windings through the current transducer to determine the actual wattage. Alternatively, an ammeter may be used in the above procedure to measure amperage instead of wattage; multiply the measured amperage by the system voltage and the assumed power factor to determine wattage (Watts = Volts × Amps × Power Factor).

Measuring Hours of Operation

One of the most critical assumptions in energy calculations is the assumption of lighting equipment operating hours. Although conservative assumptions of operating hours can be based on hours of occupancy, direct measurement has shown that the *actual* hours of system operation can far exceed these initial estimates.

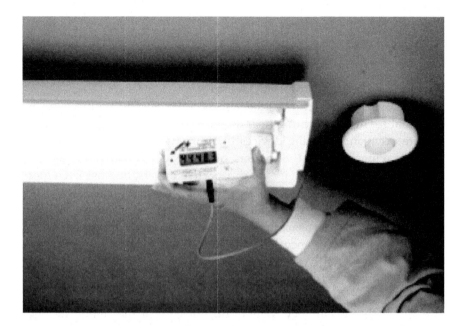

Figure 18-5. The risk in lighting upgrade investments can be minimized by performing some simple measurements. The hours of lighting operation can be measured with a logging device that senses when the lights are on. When connected to an occupancy sensor, this device counts the number of hours that the lighting system has been left on when the space was unoccupied. *Courtesy: Pacific Science & Technology.*

To measure lighting circuit operating hours, install a battery-powered "runtime logger," which mounts magnetically inside a luminaire. These devices detect whether the luminaire is on or off by sensing light. Most runtime loggers simply display the total hours of ON-time. More sophisticated devices can provide lighting system ON-time by time of use. In addition, loggers with remote sensors can be used for measuring the run time of high-temperature HID and compact luminaires.

An estimate of energy consumption can be calculated by multiplying the hours of ON-time by the fixture wattage measured previously. Where time-of-use rates are in effect, multiply the measured wattage by the hours of use within each time period defined by the rate structure. Time-of-use loggers are sold with downloading cables and computer software to facilitate data analysis.

Note that a single runtime logger reading is only valid for the circuit to which the luminaire is connected. Multiple logger readings on different circuits may be needed to obtain a more accurate picture of lighting operating hours.

Measuring Occupancy Sensor Savings

Before making an investment in occupancy sensors, measure the potential savings in lighting operating hours. Several companies manufacture occupancy logging devices that directly measure the occupancy sensor savings potential. Suppliers of occupancy sensors are eager to loan, rent or sell these devices to prospective buyers, because such measurements are often required before purchase decisions are made.

Occupancy loggers measure the number of hours that the lighting system is on while the space is unoccupied—yielding the most important number used in occupancy sensor savings calculations: savings in lighting operating hours. These devices also take into account the time-delay period between the last motion sensed and when the sensor would turn off the lights.

Note that occupancy loggers do not control the lights. Instead, these devices are measuring the *potential* savings, based on the measured occupancy patterns and manual switching behavior. Therefore, do not announce to the occupants in the test spaces that an occupancy logger has been installed to measure potential savings. With such information, occupants may modify their behavior and suddenly become more diligent about turning off the lights when they leave the room.

Measuring Lighting Circuit Energy Use

One alternative to lighting loggers is to directly meter the branch circuit's energy consumption in kilowatt-hours. This approach can only be used on branch circuits dedicated to lighting loads, as is common with 277V lighting systems. Although this approach is less expensive than simultaneously measuring multiple switching circuits with lighting loggers, it does not yield information about lighting usage in individual spaces. Energy meters are available for submetering branch circuit energy use in kilowatt-hours.

Measuring Savings From Dimming Controls

Many cost-effective opportunities for using automatic dimming controls have been missed because of the perception that it is difficult to determine energy savings from a variable-wattage load. Although the savings may be difficult to forecast strictly through engineering calculations, it is easy to *measure* the savings achieved by a dimming system trial installation:

1.   Measure the *kilowatts of the static load* (base case) before the upgrade. The kilowatts—$(kW)_{before}$—should represent the power drawn by all the fixtures on the test circuit when the switch is turned on.

2.   Install the dimming control system, and include a runtime logger in one of the fixtures. In addition, install a kilowatt meter on the lighting circuit.

3.   During a test period of several weeks, allow the new lighting system to operate with the automatic dimming control. At the end of the test period, record the number of hours of lighting system "on" time as indicated by the runtime logger—$(hrs)_{after}$. And for the same test period, record the number of kilowatt-hours consumed by the lighting system—$(kWh)_{after}$.

4.   To determine the actual energy savings achieved by the dimming control system, plug the values recorded above into the equation below.

$$kWh_{saved} = [(kW)_{before} \times (hrs)_{after}] - (kWh)_{after}$$

Note that if the installation of the dimming control system included a conversion to electronic ballasts, the savings calculated above will include the savings due to the more efficient ballasts. Therefore, to determine the savings due strictly to the dimming control, start by measuring the static wattage of the dimmable system at full light output.

EVALUATING OCCUPANT ACCEPTANCE

When planning the installation of a lighting upgrade project, it is essential to consider how the occupants will accept the proposed changes. Humans are known to first react negatively to change, unless they are properly informed about the change. Compounding this phenomenon is that most occupants assume that "energy conservation" is synonymous with "doing more with less." If, however, the occupants are shown that the new lighting system will help them do their job better, improve aesthetics, or simply improve the environment, they will be more inclined to favor the change.

Employee Education

Employee education helps boost acceptance of new lighting systems when they realize that their company is concerned about employee visual comfort and productivity as well as improving job security through reduced waste and improved competitiveness. Employees should understand that the primary goal of the lighting upgrade is to enhance working conditions such that productivity would be improved, given that the cost of labor can be worth over 200 times more than energy savings. What most employees may not understand at first is that most lighting quality improvements can be achieved while improving efficiency and saving energy.

Occupant Acceptance through Trial Installations

Install a trial installation (in a willing employee's work space) that will demonstrate the efficiency and quality improvements resulting from the lighting upgrade project under consideration. Employees will be better able to appreciate the quality improvements if the trial installations are made immediately adjacent to a similar space that has not been upgraded.

Trial installations are particularly useful for allowing employees to experiment with new technologies before these technologies are permanently installed. Some lighting upgrades that maximize energy savings may require additional demonstration time. These upgrades include task/ambient lighting, occupancy sensors and dimming controls. Once the occupants have had a chance to evaluate the reliability and flexibility of these upgrades for meeting their needs in a demonstration installation, they will be more inclined to accept them when installed in their own space.

To further enhance occupant acceptance, quantify the savings that the upgrade will yield in terms of energy reduction, cost avoidance, pollution prevention, products sold, or even jobs saved! Publicize this information along with a listing of the visible quality improvements such as visual comfort and color rendering. Occupants should be encouraged to visit the trial installation to evaluate the improvements and understand the impacts the lighting upgrade would have if installed throughout the facility.

Chapter 19

Project Implementation

Once the lighting upgrade opportunities have been assessed, the work involved in project implementation begins. This chapter guides facility managers through the various implementation steps, including 1) identifying project implementation resources, 2) evaluating financing options, 3) convincing landlords/tenants, 4) negotiating purchasing agreements, 5) managing the project and 6) commissioning the new system.

PROJECT IMPLEMENTATION RESOURCES

Although many companies have the resources to implement lighting upgrades using in-house staff exclusively, most lighting upgrade projects will require the assistance of outside professionals to perform specific tasks. Table 19-1 shows the types of expertise needed and the available contact sources. The contact sources can be reached using the information provided in Chapter 21.

EVALUATING FINANCING OPTIONS

Most comprehensive lighting upgrade projects cost between $0.50 and $2.00 per sq.ft. of floorspace. Although most companies would choose to use internal funds for financing lighting upgrades, some upgrades may require outside sources of capital. Even when factoring in the interest expense associated with borrowing money, lighting upgrades can be an excellent investment. Perhaps the most compelling reason to consider outside financing sources is that the periodic financing payments can be structured to be less than the periodic energy savings, resulting in *immediate positive cash flow.*

Table 19-1. Sources for lighting upgrade expertise.

Area of Expertise	Contact Sources
Product Suppliers	• IESNA Lighting Equipment & Accessories Directory • Energy User News Product Directories • National Electrical Manufacturers Association (NEMA)
Surveyors and Consultants	• IESNA Membership • International Association of Lighting Designers (IALD)
Installation Contractors	• interNational Association of Lighting Management Companies (NALMCO)
Maintenance Contractors	• interNational Association of Lighting Management Companies (NALMCO)

Private Sector Financing Options

Profitable, tax-paying entities in the private sector generally use the more conventional financing methods outlined in Table 19-2.

Conventional Loans

Conventional lending institutions may provide the needed capital for procuring lighting upgrades. Investigate sources where the owner may already have a credit history or where an existing credit relationship exists.

Capital Leases

Capital leases are structured very much like conventional loans. They are installment purchases that require little or no initial capital outlay, which can have a positive effect on profits as measured by net present value. The building owner is considered the owner of the equipment and may take deductions for depreciation and for the interest portion of payments to the lessor. Capital leases are offered by banks, leasing companies, installation contractors, suppliers and some electric utilities.

Table 19-2. Private sector lighting upgrade financing options. *Source: EPA Green Lights.*

	Cash Purchase	*Conventional Loans*	*Capital Lease*
Initial Payment	100% of project cost	0-30% of project cost	$0 or deposit
Periodic Payments	none	fixed	fixed
Payment Source	capital	capital	capital
Performance Risk[1]	owner 100%	owner 100%	owner 100%
Contract Termination Options	n/a	principal payoff	principal payoff
Upgrade Ownership	building owner	building owner	building owner
Tax Deductions[2]	depreciation	depreciation and interest	depreciation and interest

[1]Owner's risk may be reduced with guaranteed savings insurance.
[2]Subject to changes in tax laws. Consult with a tax advisor regarding eligibility.

Note that when using either bank financing or capital leasing arrangements, the financed amount will be recorded as debt on the participant's balance sheet. There are other options that are considered as "off-balance-sheet" financing (such as operating leases and fixed-term shared savings), but they may violate financial accounting standards when used strictly with lighting equipment. This is because most lighting equipment does not have sufficient resale value at the end of the financing term to meet the requirements of the Financial Accounting Standards Board Statement 13 (FASB 13). In most cases, off-balance-sheet financing will only be allowed in the private sector when used to finance equipment with a higher percentage of resale value (such as furniture, cars and other "rentable" equipment, such as chillers and other capital equipment).

Public Sector Financing Options

Public entities usually have several options to consider for financing their lighting upgrades. Although using internal budgets can work in some cases, there are many situations where the budgets are too tight, the budget cycle is too long, or the budget shrinks or changes over time. The financing options that public sector organizations may consider are outlined in Table 19-3.

Table 19-3. Public sector lighting upgrade financing options. *Source: EPA Green Lights.*

	Cash Purchase	Municipal Lease	Fixed-Term Shared Savings	Variable-Term Shared Savings
Initial Payment	100% of project cost	0-30% of project cost	$0	$0
Periodic Payments	none	fixed	variable	fixed
Payment Source	capital	operations	operations	operations
Performance Risk[1]	owner 100%	owner 100%	investor 100%	shared risk
Contract Termination Options	n/a	principal payoff	fair market value buyout, renewal	principal payoff
Upgrade Ownership	building owner	building owner	investor	building owner

[1]Owner's risk may be reduced with guaranteed savings insurance.

Municipal Leasing

One popular method for public entities to consider for project financing is the municipal lease. This is a form of bond financing, but it does not require a public vote. Municipal lease rates are low because the investors receive tax-free interest income. There is usually no down payment, and in some cases, the first payment can be deferred for a few months. Otherwise, the financing works somewhat like a conditional sales agreement, with payments based on standard principal and interest calculations.

Fixed-term Shared Savings

Fixed-term (or true) shared savings is one of two forms of shared savings used by public entities. A third-party investor provides the capital for the project and bears the performance risk of the investment. The public entity makes payments based on a fixed percentage of the measured energy savings. The payment amounts can vary, just as measured energy savings can vary from month to month. During the fixed term, the investor/provider owns and maintains the equipment in such a way to ensure

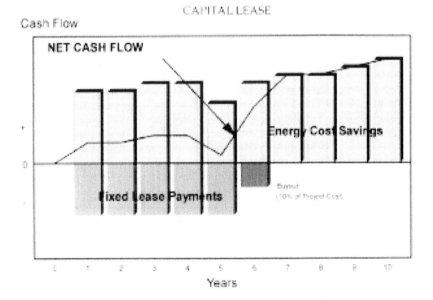

Figure 19-1. Representative cash flow when purchasing lighting upgrades with a capital lease. Note that the payments are structured to provide positive cash flow. *Courtesy: EPA Green Lights.*

that savings continue. At the end of the contract term, the entity can purchase the equipment at fair market value, or renew the agreement for an additional term (while negotiating a reduced payment percentage).

This form of financing is rarely used for lighting because of the relatively high cost of providing ongoing savings verification. Usually, this form of financing is used when the risk of the investment is relatively high, and the scope of the project addresses multiple end uses of energy (HVAC, hot water, etc.). When comparing the various financing options, shared savings is usually the most expensive on a net present value basis. However, if the entity is unable to incur additional debt, this may be one of the few options available.

Variable-Term Shared Savings

Another form of shared savings results in a sharing of risk. In variable-term shared savings, payments are based on a fixed percentage of

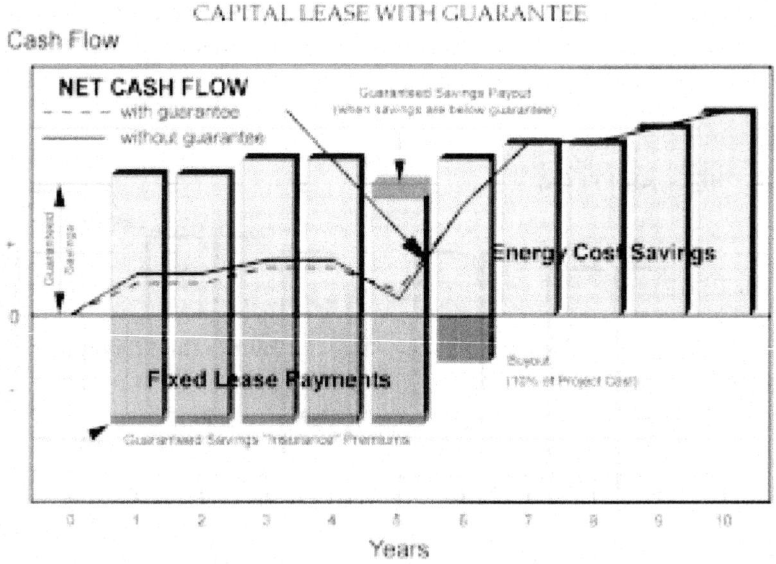

Figure 19-2. Representative cash flow when purchasing lighting upgrades with a capital lease and including a guarantee that the annual energy cost savings will exceed the annual lease payments. Note that when the total lease payments in year 5 exceeded the energy cost savings, the guarantee provided the difference in cash. *Courtesy: EPA Green Lights.*

measured energy savings, but the equipment ownership remains with the public entity. The financing works like a loan, but the payments and the financing term can vary. If energy savings increase, payments will also increase, which will cause a reduction in the financing term (a faster payoff). If the savings are less than expected, the project takes longer to pay off—up to a maximum term. There are various contract termination options.

Guaranteed Savings Insurance

In spite of the low-risk nature of lighting investments, some decision-makers in both private-sector and public-sector organizations will need to have a guarantee of cost savings performance in order to approve the project or secure financing. Most guarantees will insure that the actual measured savings will exceed a fixed percentage of estimated energy savings. If the project produces savings that are below this minimum, then the insurance provider will make up the difference.

Like any insurance, the greater the coverage (or risk mitigation), the higher the cost of insurance. Therefore, the insurance premiums for guaranteeing 90 percent of the estimated savings will cost more than guaranteeing 70 percent of the estimated savings.

Although this performance insurance can be applied to any type of procurement method (including internal funding), guaranteed savings programs are usually linked to a financing arrangement to guarantee that the periodic savings will be greater than the periodic payments, thereby *guaranteeing positive cash flow*. Again, like other insurance policies, users will be sacrificing some savings (in the form of premium payments) in order to shift some of the performance risk to the provider. The premiums also cover the cost of periodically measuring savings.

CONVINCING LANDLORDS AND TENANTS

For decades, tenants and landlords have engaged in leasing practices that create disincentives to invest in projects that reduce operating expenses. Net leases require the tenant to pay for electricity costs, typically using an expense-allocation formula. If there is no mechanism for allowing the landlord to receive energy cost savings, there is no incentive for the landlord's investment.

However, net leases *can* be renegotiated to lower the tenants' occupancy costs *and* enable the landlord to earn a fair return on the lighting upgrade investment. As illustrated in Table 19-4, this goal is accomplished by raising the tenants' rent portion of their monthly payments by *less* than the reduction in the electricity expense portion of their monthly payments.

This modified lease will allow the tenants to enjoy improved lighting systems and reduced monthly costs without making any investment. The landlord's investment increases the building's net operating income by reducing expenses, which in turn increases the building's asset value. High quality lighting upgrades in income properties can provide a marketing advantage to the landlord in attracting new tenants while helping to retain existing tenants.

NEGOTIATING PURCHASING AGREEMENTS

Product costs can be lowered and supplier service improved through national purchasing agreements, also called national accounts. These are negotiated relationships between suppliers and nationwide buyers of products and services. National accounts provide these benefits:

- Streamline coordination of lighting equipment purchases.

- Guarantee the availability of selected technologies.

Table 19-4. Example of reduced tenant occupancy costs on renegotiated net lease. *Source: EPA Green Lights.*

Before Upgrade	$/sq.ft./yr	After Upgrade	$/sq.ft./yr
Base Rent	$17.50	Base Rent	$17.71
Electricity	$ 1.75	Electricity	$ 1.40
Other Costs	$ 6.00	Other Costs	$ 6.00
Total	$25.25	Total	$25.11

- Ensure competitive prices.

- Allow for multilocation shipping direct from the manufacturer.

- Standardize installation and maintenance of the lighting equipment.

- Provide added support services.

National account agreements may or may not be written agreements. To be legally binding, the agreements are written, agreed to and signed by both parties. A Request for Proposals (RFP) to solicit bids on a national account may or may not be issued. To initiate national account relationships with suppliers, follow the guidelines presented below:

- Determine the current quantity and price for the lamps, fixtures and services purchased.

- Plan and aggregate company-wide purchases to gain the maximum discount and other benefits relating to high-volume purchasing.

- Identify which products will be specified for purchase and whether substitutes will be accepted.

- Determine projected annual purchasing volume.

- Contact the appropriate manufacturers to inquire about establishing a national account.

- Issue an RFP, if necessary, to solicit bids from interested lamp, ballast, fixture and service companies for products and services to be included in a national account agreement.

MANAGING THE PROJECT

The implementation of lighting upgrade projects requires timely and organized management of labor, material and administrative resources. This section provides a checklist of issues and tasks that are important to successful project management. The tasks involved in managing lighting upgrade projects are similar to those for managing any building systems project.

Safety and Insurance

• Bid bonds may need to be submitted with the proposal to guarantee that the bid has been made in good faith and that the bidder will enter into the contract if their proposal is accepted.

• Performance bonds may be required to guarantee that the contractor will perform the work specified in the contract in accordance with its terms.

• An agreement to arbitrate disputes may be included in the contract, if deemed necessary.

• If asbestos exists in the ceiling system, proper precautions need to be made to ensure worker safety and to comply with handling and disposal regulations.

• Verify that all products meet applicable building codes and standards. Arrange for acquiring building permits or arranging building inspections if required by the local jurisdiction.

• Make sure that contractors carry sufficient insurance, including workman's compensation, public liability and automobile insurance. Consider asking the contractor to supply a certificate of insurance listing your company as additionally insured.

Project Start-up

• Review the lighting upgrade specifications to verify that they meet the organization's requirements.

• Request product samples and arrange trial installations to evaluate alternative technology options prior to full-scale installation.

• Schedule periodic project meetings to review progress to date and resolve any problems.

• Discuss conditions for approving subcontractors, as needed.

• Agree on locations where equipment and materials can be stored.

• Check hazardous waste regulations for maximum allowable storage periods and proper handling, transportation and disposal of lighting waste (see Chapter 20).

- Designate contractor parking areas as needed.

- Agree on the schedule of work, addressing both hours and days.

- Arrange for contractor access to the job site. Initiate key control measures as required.

Project Installation
- Establish minimum performance expectations for daily clean-up and waste removal.

- Set standards for project supervision.

- Consider how changes in the scope of work will be negotiated and compensated.

- Establish the procedure for approving change orders.

- Verify that the contract is clear regarding payment schedules and conditions.

- Establish standards and procedures for quality control.

- Verify that each lighting upgrade product is furnished with a warranty.

- Negotiate for extra stock of new components (lamps, ballasts, etc.) to be furnished at project completion for future maintenance needs.

COMMISSIONING THE NEW SYSTEM

Following the completion of the lighting upgrade installation, take steps to ensure that the new lighting systems will continue to operate correctly and that maintenance procedures are implemented to minimize system "down time" in the event of a premature component failure. Maintaining the quality of the lighting systems will have a positive effect on sustaining worker productivity and minimizing long-term energy costs. The aspects of effective system commissioning are described below.

Contractor Support

Included as part of the scope of work under the installation contract, the contractor should provide 90 days of post-occupancy services as needed to ensure that the commissioning requirements described below are met. Examples of these contractor-provided services include replacement of any defective component, warranty applications, adjustments to occupancy sensors and calibration of daylight controls. In addition, the contractor should provide training to the building maintenance staff to enable them to perform all applicable operation and maintenance functions.

Purchase Of Replacement Stock

To minimize down time between component failure and replacement, the building staff should maintain a minimum stock of replacement components. Items to keep in stock may include lamps (all applicable types), ballasts, replacement louvers or lenses, occupancy sensors, photosensors and task lights. The volume of inventory depends on the expected rate of failure or turnover, which depends on equipment lifetime as well as the chosen lighting maintenance strategy. Larger lamp inventories are needed when lighting systems are relamped on a "spot" basis instead of a "group" basis (see Chapter 20 for more on lighting maintenance strategies).

Fixture Labeling

In applications where fluorescent luminaires are tandem-wired with 4-lamp electronic ballasts, *every other* luminaire actually contains a ballast. Those luminaires that house a ballast should be identified with a small adhesive label on the trim or interior of the luminaire. Similarly, if both partial-output and full-output electronic ballasts are used in the same facility, a color-code labeling strategy will help the maintenance staff identify and install the correct ballast (and maintain the energy savings and proper light levels).

Ongoing Maintenance Staff Training

As new maintenance staff members are hired, they should receive training on the system maintenance procedures as identified by the contractor or supplier. In addition, product literature, equipment specifications, as-built floor plans, equipment sources and key contact information should be maintained in a document for convenient future reference.

Chapter 20

Lighting Maintenance

From the moment a lighting system upgrade is installed, the light output will begin to decline. Most occupants do not notice the gradual decline in light levels because their eyes adapt to gradually changing lighting conditions. But eventually, the reduction may affect the appearance of the space as well as the productivity and safety of the occupants. Planned maintenance can improve light levels and offer energy-saving opportunities.

INCREASING EFFICIENCY
THROUGH IMPROVED MAINTENANCE

Over time, *all* lighting systems lose efficiency. Because fixture wattage remains relatively constant over time, this loss in efficiency is due to the gradual reduction in system light output over time. The principal causes for losses in system light output are *lamp lumen depreciation* (aging lamps) and *luminaire dirt depreciation* (dirt accumulation). Through improved maintenance practices, however, the effects of these factors can be minimized.

When planning lighting upgrades, the recommended light levels are based on *maintained* footcandles, which take into account the effects of lamp lumen depreciation and luminaire dirt depreciation. The overall effect of these factors can be significant because they are *multiplied* together to determine the average lighting system performance over time. Therefore, to maintain high performance, lighting systems should be cleaned and relamped on a regular basis.

Lamp Lumen Depreciation

As a lamp is used, the amount of light it produces declines. This phenomenon is called lamp lumen depreciation (LLD). LLD can be caused by several factors, such as carbon deposits inside the lamp wall or deterioration of the phosphor coating inside the lamp. The rates of lamp lumen depreciation for several common lamp types are shown in Figure 20-1.

In lighting calculations, the rate of LLD is expressed as a decimal value which represents the percentage of initial lumen output that the lamp produces when it reaches 40 percent of its rated life. For example, if a T8 lamp is rated at 20,000 hours, has an initial lumen output of 2,850 lumens, and has an LLD factor of 0.91, then after 8,000 hours of lamp operation (40% of 20,000), the lumen output will have dropped to approximately 2,594 lumens (0.91 x 2,850). This lumen output represents the *average* lamp output over its 20,000-hour life. Common factors for LLD are shown in Table 20-1.

Table 20-1. Typical lamp lumen depreciation factors.

Lamp Type		LLD (% of Initial Lumens @ 40% Lamp Life)
F40T12	Cool White (62 CRI)	0.87
	73 CRI and 85 CRI	0.90
F32T8	75 CRI	0.91
	85 CRI	0.93
F96T12 Slimline	Cool White	0.88
	73 CRI and 85 CRI	0.94
F96T12 HO	Cool White	0.83
	73 CRI and 85 CRI	0.90
F96T12 VHO	Cool White	0.75
F96T8 Slimline	75 CRI and 85 CRI	0.91
F96T8 HO	85 CRI	0.90
T5		0.95
T5 High Output		0.95
Compact Fluorescent		0.85
Mercury Vapor		0.79
Metal Halide		0.83
High-pressure Sodium		0.91

Luminaire Dirt Depreciation

As dirt, dust and smoke film accumulate on luminaire surfaces, light output and luminaire efficiency continually decline over time. Eventually, it becomes economical to expend the resources to clean the luminaires and restore the illumination that was lost to luminaire dirt depreciation (LDD).

The effect of dirt accumulation on luminaire efficiency is dependent on three variables:

- The dirt conditions within the room.

- The type of luminaire in use.

- The frequency of luminaire cleaning.

Dirt Conditions

The IESNA has defined five levels of dirt conditions as shown in Table 20-2. The dirtier the environment, the more frequently the luminaires should be cleaned in order to maintain high efficiency.

Type of Luminaire

The type of luminaire in use dramatically affects the rate of LDD. For example, indirect luminaires—which provide 100 percent

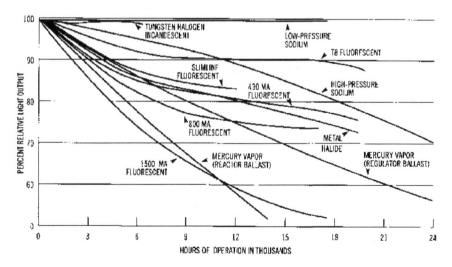

Figure 20-1. Lamp lumen depreciation curves for common light sources. *Courtesy: National Lighting Bureau.*

uplighting—will suffer the greatest LDD of all luminaire types because dirt will settle on the lamps and optical surfaces. Other luminaires that provide some degree of uplighting will also require more frequent cleaning. Refer to Table 20-3 for the various categories of luminaires with respect to luminaire dirt depreciation characteristics.

Frequency of Luminaire Cleaning

IESNA publishes rates of LLD based on the above two variables (dirt conditions and type of luminaire). After identifying the dirt conditions and luminaire type, refer to the charts in Figure 20-2 to determine the effect of LDD on light output over time. Generally, if the cleaning interval is known, use the LDD value at the end of the cleaning interval; if cleaning is not planned, use the LDD value at three years, or extrapolate the curve to the extent necessary to predict the LDD value beyond the three-year period (alternatively, refer to the IESNA *Lighting Handbook* for the equation for calculating this long-term LDD value). Luminaires should be cleaned at the same time they are group-relamped. In cases where rapid dirt depreciation is a problem, the luminaires should be cleaned more frequently.

Example: Poor Maintenance Means Poor Efficiency

The following example illustrates the impact of light loss factors on lighting system efficiency. Assume that open-fixture-rated metal halide lamps are used in a downlighting system where the dirt environment is defined as medium. In addition, assume that the lamps are replaced individually upon burnout (assuming a 20,000-hour rated lamp life) and that the luminaires are cleaned every three years. In this case, the light loss factors are:

LLD   =   0.83 (@ 40% of 20,000-hour rated life)

LDD   =   0.62 (Category IV at 36 months)

Light Loss Factor   =   LLD × LDD = 0.83 × 0.62 = 0.51

Under this maintenance scenario, the lighting system is delivering only about *half* (51 percent) of its initial light output (on average) due to the combined effects of lamp lumen depreciation and luminaire dirt depreciation.

Table 20-2. Definitions of dirt conditions. Source: *Illuminating Engineering Society of North America*.

	Generated Dirt	Ambient Dirt	Removal Or Filtration	Adhesion	Examples
Very Clean	None	None (or none enters area)	Excellent	None	High-grade offices, not near production; laboratories; clean rooms
Clean	Very little	Some (almost enters)	Better than average	Slight	Offices in older buildings or near production; light assembly; inspection
Medium	Noticeable but not heavy	Some enters area	Poorer than average	Enough to be visible after some months	Mill offices; paper processing; light machining
Dirty	Accumulates rapidly	Large amount enters area	Only fans or blowers if any	High; probably due to oil humidity or static	Heat treating; high speed printing; rubber processing
Very Dirty	Constant accumulation	Almost no dirt excluded	None	High	Similar to "Dirty" but luminaires within immediate area of contamination

Example: Improved Maintenance Means Increased Efficiency

Using the same lighting system as described in the above example, now assume that improved maintenance practices are implemented. Assume that the lamps are operated 6,000 hours per year and group-replaced every two years, at approximately 60 percent of the metal halide lamps' 20,000-hour rated life. Also assume that the luminaires are cleaned at the same time—every two years. Now, the light loss factors are:

LLD   = 0.88 (@ 40% of the 12,000-hour relamping interval)

LDD   = 0.69 (24-month cleaning interval, Category IV)

Light
Loss Factor = LLD × LDD = 0.88 × 0.69 = 0.61

With improved maintenance, the lighting system operates at 61 percent of initial light output, on average. *This 20 percent increase in system efficiency (light output) can be translated into energy cost savings by using reduced-output lamps.* Proper maintenance planning, however, is essential for sustaining this improved level of performance.

Table 20-3. Luminaire maintenance categories. *Adapted from the IESNA.*

Luminaire Maintenance Category	Representative Luminaire Types
I	Industrial strip without reflector (no top or bottom enclosure)
II	Direct-Indirect luminaire (at least 15% uplighting through apertures; remainder of lighting through bottom apertures)
III	Industrial strip with apertured reflector (1-15% uplighting through apertures; remainder of lighting through bottom aperture[s])
IV	Deep-cell parabolic troffer; open HID luminaire (top enclosure, downlighting through bottom aperture[s])
V	Lensed fluorescent troffer; enclosed HID luminaire (top and bottom are enclosed)
VI	Indirect fluorescent (100% uplighting through aperture[s]; bottom is enclosed)

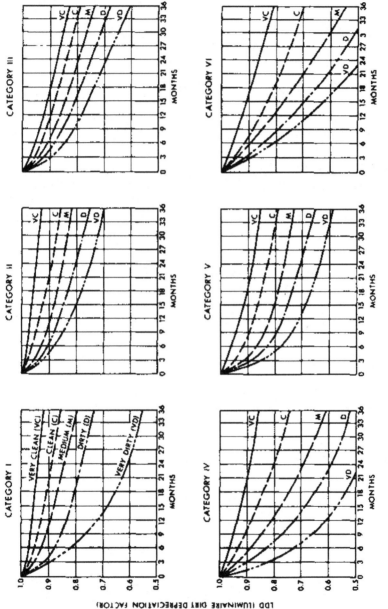

Figure 20-2. Luminaire dirt depreciation curves for the six IESNA luminaire maintenance categories. *Courtesy: Illuminating Engineering Society of North America, New York City.*

SAVING MONEY THROUGH
IMPROVED MAINTENANCE

In addition to improving system efficiency and light output, improved maintenance practices can reduce expenses. By switching to a "group maintenance" strategy, labor costs associated with fixture relamping and cleaning can be cut by over 70 percent. And the expense of lamp purchases can be reduced through fewer, higher volume transactions. The bottom line: With improved maintenance practices, building managers will get more lighting value per dollar expended.

Group Relamping and Cleaning

Instead of replacing lamps one at a time as they fail, consider the gain in labor efficiency that results when all the lamps are replaced at the same time according to a regular schedule—*before* they fail. And cleaning fixtures at the time of group-relamping can keep them operating at high efficiency.

A group relamping program can be cost-justified by the resulting improvements in labor productivity. Although procedures can vary from job to job, here is how a team of two workers can efficiently wash and relamp a fluorescent luminaire: The person on the ladder removes the lamp shielding assembly and the used lamps and hands them to the other person working on the floor below. The "floor" person unpacks the new lamps, packages the used lamps, rinses the washing sponge in the cleaning solution and cleans the lens or louver. The "ladder" person cleans the inside of the luminaire and installs the new lamps.

Many maintenance managers are hesitant to replace lamps that are still operating. But after comparing the average annual cost of sporadic spot maintenance to that of group maintenance, many have decided to switch to a group maintenance strategy. Note that in Figure 20-3, the slight increases in lamp material and disposal costs are usually far outweighed by the savings in lamp replacement and cleaning costs. Lighting management companies can assist in determining the site-specific labor cost savings that can be achieved by implementing a group maintenance program.

With an understanding of the types of lamps in use and their operating schedule, it is possible to predict lamp failure rates so that all lamps can be replaced just before frequent failures begin. To determine the optimal time to group relamp, refer to the lamp's mortality curve

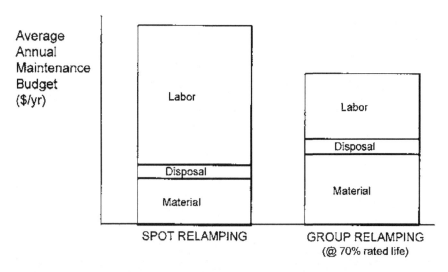

Figure 20-3. Due to the significant savings in lamp replacement labor costs, group relamping can minimize lighting maintenance costs. *Courtesy: EPA Green Lights.*

which represents the cumulative percentage of lamp failures that occur over time. Note that in Figure 20-4, the rate of fluorescent lamp failures becomes significant at approximately 70 percent of rated life (at the "knee" of the curve). HID relamping intervals are typically scheduled between 60 and 70 percent of rated lamp life, depending on the cost of spot-relamping failures that inevitably occur before group relamping is performed.

Another method for scheduling group relamping involves careful recordkeeping of spot relamping due to early failures. If relamping records indicate that 10 percent of the lamps in a previously group-relamped system have been spot-replaced due to early lamp failures, then it may be time to group-relamp the entire system.

Regardless of the scheduling method chosen, the group replacement of lamps will reduce the light loss caused by lamp failure and will reduce the time, effort and complaints associated with spot replacement of lamps. The few lamps that fail between group replacements can be tolerated or spot-replaced as needed.

Where Group Maintenance Makes the Most Sense

Although there are economies of scale associated with any group

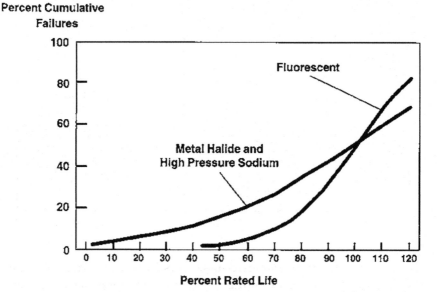

Figure 20-4. Typical lamp mortality curves. Note that 100 percent rated life is defined when 50 percent of the lamps in a large sample have failed. *Courtesy: EPA Green Lights.*

maintenance strategy, there are certain situations where group relamping and cleaning makes the *most* sense. These situations include:

- High mounting heights, where the costs of scaffolding and business disruption can be exorbitant.

- Dirty environments and/or indirect luminaires, where the rate of LDD is high.

- VHO fluorescent systems, where rapid LLD quickly reduces system efficacy.

- Metal halide systems, to minimize lamp-to-lamp color differences and to minimize the possibility of "non-passive" failure (lamp explosion) at end of life.

- Uniform hours of operation, where most lighting systems are operated on the same schedule; systems controlled with occupancy sensors create less predictable operating schedules and, therefore, less predictable lamp failure rates.

- Retail establishments, where the aesthetic impact of expired lamps creates a negative impression in the buyer's mind.

MAINTENANCE PLANNING

To derive the benefits of improved maintenance strategies, the building owner must be committed to a specific lighting maintenance plan. Otherwise, the maintenance approach will nearly always revert to inefficient "spot" relamping.

Budgeting for Group Maintenance

Budgeting is a difficult part of planning a maintenance program. Although spot maintenance of a lighting system can be sporadic on a daily basis, the annual cost will be fairly constant after the first few years. On the other hand, strategic maintenance is easier to manage on a daily basis and may cost less overall, but the total cost fluctuates each year. For example, if a lighting system with 20,000-hour lamps is operated 4,000 hours/year, then a spot-maintenance program would end up replacing about 20 percent of the lamps (at random) every year. However, maintaining the same facility on a group basis would require very few replacements for 3 years, and then 100 percent lamp replacement every forth year.

Because many budgets are established a year in advance, it may be necessary to predict the relamping schedule and budget accordingly. As a more practical alternative, lighting maintenance budgets can be leveled by completing an equal portion of the group maintenance program each year. In the example above, group-relamping 25 percent of the facility each year will keep the annual costs relatively even.

Writing a Lighting Maintenance Policy

After establishing a budget, write a lighting maintenance policy. This will assist in gaining policy approval, and it provides direction to new maintenance personnel who join the organization. In addition, the policy may be used for planning lighting maintenance procedures for other similar facilities. Although the primary content of the policy is a listing of procedures, it should also include the cost justification for the maintenance plan so that future managers will be motivated to continue the policy.

The following questions should be addressed when planning a lighting maintenance strategy:

Use Existing Staff or Use a Contractor?

In many cases, the in-house labor savings from group maintenance will free up maintenance staff to be utilized for other important tasks. If a lighting management contractor is used, even more in-house maintenance time will be freed up. Lighting management companies have developed specialized skills for quickly and effectively completing "wash and relamp" jobs; their speed and expertise can justify their higher labor rates when compared to using in-house labor. In addition, they can help develop a planned lighting maintenance program as well as provide luminaire repair services as needed. Directories of lighting management companies are available from the interNational Association of Lighting Management Companies (NALMCO).

Complete Maintenance During Regular Hours, Nights or Weekends?

It is rarely a good idea to perform lighting maintenance during regular hours. One of the benefits of a group relamping program is that most lamp replacements can be deferred to after hours, when occupants will not be disturbed by the work. Some lighting management companies may charge the same labor rate, regardless of the time of day.

How Will Quality Control Be Managed?

Issues to consider regarding quality control include safety precautions, thoroughness of luminaire cleaning, care for occupant furniture and equipment, jobsite cleanup, proper disposal of lighting waste and lighting system repairs (such as replacement of failed ballasts and cracked lenses). Job inspection guidelines should be delineated in the policy.

How Many Fixtures Will Be Relamped at One Time?

As mentioned above, it may be practical to relamp a portion of the building each year in order to level out the annual maintenance expenditures. In some cases, it may be more feasible to limit the group-relamping to even smaller sections. However, depending on the size of the system, the types of light sources and their operating schedules, it may be more economical to relamp the entire building at once.

What Procedures Will Be Followed for
Testing Emergency Lighting Systems?
The emergency lighting systems must be tested periodically as required by the state or local code. In addition, state or local codes may require that periodic maintenance on battery systems be performed and documented. Procedures may involve checking and refilling battery water level, testing battery voltage in stand-by mode, testing lighting in backup power mode for five minutes, rechecking battery voltage after the 5-minute test discharge, checking for proper operation of the battery charging system, and entering documentation of maintenance and testing in a permanent file.

How Will the Lighting Waste Be Disposed?
Refer to the following section regarding the best practices for disposing of used lamps and ballasts.

LIGHTING WASTE DISPOSAL

As lamps and ballasts are removed from service during lighting upgrades or routine maintenance, building owners are responsible for properly disposing of this waste. There are two types of potentially hazardous lighting waste: mercury-containing *lamps* (fluorescent and HID lamps) and PCB-containing *ballasts*.

Although there are a number of confusing regulations that affect disposal of lighting waste, it is easiest and nearly always cost-effective to simply recycle the waste. The only limitation to this general rule is that *leaking* PCB ballasts cannot be recycled (they must be incinerated). Because regulations vary from state to state and are subject to change, contact the applicable state solid and hazardous waste agency.

Disposing of Mercury-containing Lamps
All fluorescent and high-intensity discharge lamps contain a small quantity of mercury that can potentially be harmful to the environment and human health. Because the costs of discarding lamps by recycling or hazardous waste landfill are very low compared to lamp material and energy costs, the additional costs of proper lamp disposal rarely cause lighting upgrades to become unprofitable.

Federal Regulations

Currently, the U.S. Environmental Protection Agency (EPA) regulates mercury disposal under the Resource Conservation and Recovery Act (RCRA). A testing procedure known as the Toxic Characteristic Leaching Procedure (TCLP) identifies whether a waste is toxic and must be managed as hazardous waste. When lamps are tested using the TCLP, the results can vary considerably depending on the lamp manufacturer's dosing techniques, the hours of lamp use before disposal and the laboratory procedures used. Because of the high cost of testing a lamp, and because the results of one lamp test do not apply to other lamps, it is best to consider all lamps as hazardous waste and dispose of them accordingly.

Exemption for Small Quantity Generators

Companies that generate less than 100 kg of hazardous waste per month are excused from RCRA regulations regarding hazardous wastes. If all 100 kg of the waste are lamps, this exemption translates to about 300-350 4 ft. T12 lamps or about 400-450 4 ft. T8 lamps per month.

Disposal Options

Used fluorescent and HID lamps may be recycled or disposed of in hazardous waste (Subtitle C) landfills. Mercury-containing lamps should *not* be incinerated because the elemental mercury cannot be destroyed, and the airborne emissions pose a greater threat to the environment and human health than landfill disposal. The cost for recycling fluorescent lamps is about 50-75 cents per 4 ft. fluorescent lamp, while hazardous

---

LOW-MERCURY LAMPS

Major lamp manufacturers offer fluorescent and HID lamps that are specifically designed to pass EPA's TCLP test at end of life, potentially allowing users to dispose of these new lamps as standard municipal waste. Using improved technology, these lamps are manufactured with reduced mercury content. Although mercury is essential for the operation of all fluorescent and HID lamps, manufacturers claim that the reduction in mercury content has minimal impact on lamp life or efficacy. Check with the applicable state hazardous waste agency to determine if these lamps qualify for disposal in municipal solid waste landfills.

waste landfill costs are about 25-50 cents per 4 ft. lamp, excluding packaging, transportation and profile fees.

Disposal of PCB-containing Ballasts

Although PCB (polychlorinated biphenyl) fluids have not been used in fluorescent ballast capacitors since the late 1970s, some of these ballasts are still in operation. The primary concerns regarding disposal of PCB-containing ballasts are the health risks of direct contact with toxic PCB fluids.

Identifying PCB Ballasts

All ballasts manufactured after 1979 that do not contain PCBs are labeled, "NO PCBs." Therefore, if a ballast is not labeled "NO PCBs," assume that it does contain PCBs. If a ballast appears to be leaking PCB fluid, take precautions to prevent exposure to the leaking ballast, because all materials that contact the ballast or the leaking substance are also considered as PCB waste and must be incinerated along with the ballast. Use trained personnel or a waste management contractor to handle and dispose of leaking PCB-containing ballasts.

Federal Regulations

The federal government regulates the disposal of *leaking* PCB ballasts under the Toxic Substances Control Act

Figure 20-5. Fluorescent lamp recycling allows for the reuse of the metal, glass, phosphor powder and mercury. In addition to linear fluorescent tubes, other lamps can be recycled, including compact fluorescents, U-lamps and HID lamps. *Courtesy: Advanced Environmental Recycling Corp.*

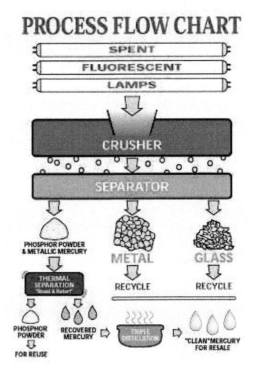

PROCESS FLOW CHART

SPENT
FLUORESCENT
LAMPS

CRUSHER

SEPARATOR

PHOSPHOR POWDER
& METALLIC MERCURY

METAL

GLASS

THERMAL
SEPARATION
"Roast & Retort"

RECYCLE

RECYCLE

PHOSPHOR
POWDER

FOR REUSE

RECOVERED
MERCURY

TRIPLE
DISTILLATION

"CLEAN" MERCURY
FOR RESALE

The EPA Universal Waste Rule

In 1999, EPA enacted a regulation that allows generators to manage lamps under a new set of modified regulations known as the Universal Waste rule. This rule significantly reduces the transportation, storage, collection and record keeping requirements for spent lamps that are *recycled*. Still, hazardous waste regulations pertaining to spent fluorescent lamps continue to vary from state to state. Facility managers should contact their state authorities to determine the spent lamp management requirements applicable to them. Information about state lamp management requirements can be obtained at *www.lamprecycle.org*.

(TSCA). Leaking PCB ballasts must be disposed of in an EPA-approved high-temperature incinerator. State and/or local regulations may apply to non-leaking PCB ballasts.

Even if fluorescent ballasts are not leaking, there may be federal regulations that address their ultimate disposal. Superfund regulations require building owners who release a pound or more of PCBs (roughly equivalent to 12-16 magnetic F40 ballasts) in a 24-hour period into the environment to notify the National Response Center at (800) 424-8802. Generators that dispose of PCB-containing ballasts at municipal, hazardous or chemical waste landfill sites may become liable for subsequent Superfund cleanup costs *even if PCB ballast disposal is allowed at these sites*. Changes are being considered in the laws pertaining to PCB ballast disposal; contact EPA for the current regulations.

State Regulations

State regulations may address methods for disposing of non-leaking PCB ballasts. Some states require that non-leaking PCB ballasts be handled, stored, transported and disposed of as a hazardous waste. Other states simply prohibit their disposal in municipal solid waste landfills, thereby forcing disposal by the more environmentally safe methods as described below. In addition, there are several other states with no specific regulations about the disposal of non-leaking PCB ballasts. It is the generator's responsibility (the building owner's responsibility) to investigate and follow state and local regulations regarding non-leaking PCB ballast disposal.

Disposal Options

As indicated above, there is only one option for disposing of leaking PCB ballasts: high-temperature incineration. However, there are three options for disposing of *non-leaking* PCB ballasts:

• The entire non-leaking ballast may be disposed in a *high-temperature incinerator*. High-temperature incineration is preferred by many companies because the process destroys the PCBs and permanently eliminates them from the waste stream, thereby removing the potential for future Superfund liability. This is one of the most expensive disposal options, costing over $5.00 per ballast, excluding packaging, transportation and profile fees (hazardous waste documentation costs).

• Although some state regulations allow non-leaking PCB ballasts to be disposed of in *hazardous waste (Subtitle C) landfills*, this disposal method exposes the generator of the waste to potential future Superfund liability, where cleanup costs are paid by those who contributed to the disposal site and could afford to carry a disproportionate share of the cleanup costs. It is not surprising that this is the least expensive disposal option, costing on average about $0.50 per ballast, excluding packaging, transportation and profile fees.

• *Ballast recycling* is another acceptable disposal method. Ballast recyclers remove the PCB-containing ballast (and the potting compound if contaminated) and dispose of it in a hazardous-waste landfill or high-temperature incinerator. The remaining materials, such as the copper and steel, can be reclaimed from the ballasts for use in manufacturing other products. The average recycling cost is about $3.50 per ballast, excluding packaging, transportation and profile fees.

Appendix I

Trade and Professional Associations

See the Bibliography of Sources in Appendix II for more information sources.

Association of Energy Engineers (AEE)
www.aeecenter.org

Illuminating Engineering Society of North America (IESNA)
www.iesna.org

International Association of Lighting Designers (IALD)
www.iald.org

International Association of Lighting Management Companies (NALMCO)
www.nalmco.org

Light Right Consortium
www.lightright.org

National Association of Energy Service Companies (NAESCO)
www.naesco.org

National Association of Independent Lighting Distributors (NAILD)
www.naild.org

National Electrical Contractors Association (NECA)
www.necanet.org

National Electrical Manufacturers Association (NEMA)
www.nema.org

National Lighting Bureau
www.nlb.org

FEDERAL GOVERNMENT RESOURCES

EPA Energy Star Buildings Program
www.energystar.gov

Federal Energy Management Program
www.eren.doe.gov/femp

SELECTED PERIODICALS

Architectural Lighting
www.lightforum.com

Energy User News
www.energyusernews.com

LD+A (Lighting Design and Application)
Illuminating Engineering Society
www.iesna.org

Lighting Management and Maintenance
International Association of Lighting Management Companies
www.nalmco.org

Lighting Futures
Lighting Research Center at Rensselaer Polytechnic Institute
www.lri.rpi.edu

Appendix II

Bibliography of Sources

A *dvanced Lighting Guidelines: 1993*, Electric Power Research Institute (EPRI)/California Energy Commission (CEC)/United States Department of Energy (DOE), May 1993.

ASHRAE/IES Standard 90.1-1989, American Society of Heating Refrigerating and Air-Conditioning Engineers (ASHRAE) and Illuminating Engineering Society (IES), 1989.

Barnhart, J.E., DiLouie, C. and Madonia, T. *Illuminations: A Training Textbook for Senior Lighting Technicians*, interNational Association of Lighting Management Companies (NALMCO), First Edition, Princeton Junction, NJ, 1993.

DiLouie, Craig. *Lighting Management Handbook*. Lilburn, GA: The Fairmont Press, 1993.

Electric Power Research Institute (EPRI) Lighting Brochures, Palo Alto, CA:
    High Intensity Discharge Lighting (1993), BR-101739
    Electronic Ballasts (1993), BR-101886
    Occupancy Sensors (1994), BR-100323
    Compact Fluorescent Lamps (1993), CU.2042R
    Specular Retrofit Reflectors (1992), CU.2046
    Retrofit Lighting Technologies (1993), CU.3040

Fluorescent Lamps and the Environment, National Electrical Manufacturers Association, 2001

IES Education Series, Illuminating Engineering Society of North America, New York, NY.
    ED-100: Introductory Lighting, 1993

ED-150: Intermediate Lighting, 1993

IES Lighting Handbook, 9th Edition, Illuminating Engineering Society of North America, New York, NY, 2000.

IES Lighting Ready Reference, Illuminating Engineering Society of North America, New York, NY, 1989.

IES Recommended Practices, Illuminating Engineering Society of North America, New York, NY:
  Office Lighting (RP-1), 1993
  Lighting Merchandising Areas (RP-2), 1992
  Educational Facilities Lighting (RP-3), 1988
  Library Lighting (RP-4), 1974
  Daylighting (RP-5), 1979
  Sports Lighting (RP-6), 1988
  Industrial Lighting (RP-7), 1991
  Roadway Lighting (RP-8), 1993
  Roadway Sign Lighting (RP-19), 1989
  Lighting for Parking Facilities (RP-20), 1985
  Calculation of Daylight Availability (RP-21), 1984

Lighting Fundamentals Handbook, Electric Power Research Institute, Palo Alto, CA, March 1993.

Lighting Research Center, Rensselaer Polytechnic Institute, Troy, NY.
  Lighting Listings: A Worldwide Guide to Lighting Publications, Research Organizations, Educational Opportunities And Associations, 1995.

  Light Conversations: Productivity, Comfort, Health, Safety, Aesthetics, 1994

  Fluorescent Lamp/Ballast Compatibility, 1994

  Lighting Regulation in The United States, 1992

  Lighting Controls: A Scoping Study And an Annotated Bibliography, 1992

Specifier Reports And *Lighting Answers* (see National Lighting Product Information Program)

Lighting Technology Atlas, E Source Inc., Boulder, CO, 1994.

Lighting Upgrade Manual, U.S. Environmental Protection Agency, Washington, DC, December 1995.

Lindsey, Jack L., *Applied Illumination Engineering*. Lilburn, GA: The Fairmont Press, 1991.

National Lighting Bureau Publications, Rosslyn, VA.
    Getting The Most from Your Lighting Dollar, 1982
    Performing a Lighting System Audit, 1985
    Profiting from Lighting Modernization, 1987
    Solving The Puzzle of VDT Viewing Problems, 1987
    Lighting And Human Performance, 1989
    Lighting for Safety and Security, 1990
    NEMA Guide to Lighting Controls, 1992
    NLB Guide to Industrial Lighting, 1992
    NLB Guide to Energy Efficient Lighting Systems, 1994
    NEMA Guide to Emergency Lighting, 1995

National Lighting Product Information Program, Lighting Research Center, Rensselaer Polytechnic Institute, Troy, NY.
    Specifier Reports:
    Power Reducers, 1992
    Specular Reflectors, 1992
    Occupancy Sensors, 1992
    Parking Lot Luminaires, 1993
    Screwbase Compact Fluorescent Lamp Products, 1993, 1994, 1995
    Cathode-Disconnect Ballasts, 1993
    Exit Sign Technologies, 1994, 1995
    Electronic Ballasts, 1994, 1995
    Reflector Lamps, 1994
    CFL Downlights, 1995
    Dimming Electronic Ballasts, 1995
    Lighting Answers:

T8 Fluorescent Lamps, 1993
Multilayer Polarizer Panels, 1993
Task Lighting for Offices, 1994
Dimming Systems for HID Lamps, 1994
EMI Involving Fluorescent Lighting Systems, 1995
Power Quality, 1995
Thermal Effects in 2'x4' Fluorescent Lighting Systems, 1995
T10 and T9 Fluorescent Lamps, 1995

Philips Lighting Application Guides, Philips Lighting Company, Somerset, NJ.
Office Lighting, 1992
Security Lighting, 1992
Industrial Lighting, 1991
Retail Lighting, 1991
Healthcare Lighting, 1992

Romm, Joseph J. and Browning, William D. Greening The Building And The Bottom Line: Increasing Productivity Through Energy-Efficient Design. Rocky Mountain Institute, Snowmass, CO, 1994.

Summary of Lighting Control Technologies And Strategies, U.S. Environmental Protection Agency, Washington, DC, 1993.

Tech Update, E Source Inc., Boulder, CO.
New Electronic Timers Simplify Lighting Control, 1995

Fluorescent Dimming for Energy Management: More Options, But Not Yet Mature, 1995

Market Survey: Lighting Loggers And Occupancy Sensors, 1994

LED Exit Signs: Improved Technology Leads The Way to Energy Savings, 1994

Does Low Ballast Factor Instant Start Operation Reduce Fluorescent Lamp Life?, 1993

Occupancy Sensors: Promise And Pitfalls, 1993

Ultraviolet Radiation from Electric Lighting, 1993

High Performance CFL Downlights: The Best And The Brightest, 1993

Impending U.S. Lighting Standards Will Boost Market for Halogen-Infrared Lamps, 1993

High Lumen Compact Fluorescents Boost Light Output in New Fixtures, 1992

Electronic Ballasts for Metal Halide Lamps Broaden Applicability And Improve Efficiency, 1992

Lighting Retrofits May Affect Insurance Coverage And Code Compliance, 1992

Electronic Ballasts: Developments in The U.S. Market, 1992

Technology Assessment: Energy-Efficient Commercial Lighting. Berkeley, CA: Lawrence Berkeley Laboratory, Applied Science Division, March 1989.

Thumann, Albert. *Lighting Efficiency Applications*, 2nd Edition. Lilburn, Georgia: The Fairmont Press, 1992.

Appendix III
Glossary of Terms

Ampere - The standard unit of measurement for electric current that is equal to one coulomb per second. It defines the quantity of electrons moving past a given point in a circuit over a period of time. Amp is an abbreviation.

ANSI - American National Standards Institute, a non-profit organization established to develop voluntary industry standards for various products.

Arc Tube - A tube enclosed within the outer glass envelope of a HID lamp and made of clear quartz or ceramic that contains the arc stream.

Ballast - A device used to operate fluorescent and HID lamps. The ballast provides the necessary starting voltage, while limiting and regulating the lamp current during operation.

Ballast Cycling - Undesirable condition whereby the ballast turns lamps on and off (cycles) due to the overheating of the thermal switch inside the ballast. This may be due to incorrect lamps, improper voltage being supplied, high ambient temperature around the fixture, or the early stage of ballast failure.

Ballast Efficiency Factor - The Ballast Efficiency Factor (BEF) is the Ballast Factor (see below) divided by the input power of the ballast. The higher the BEF—within the same lamp-ballast type—the more efficient the ballast.

Ballast Factor - The Ballast Factor (BF) for a specific lamp-ballast combination represents the percentage of the rated lamp lumens that will actually be produced by the combination.

Ballast Losses - Power supplied to the ballast but not transformed into energy used by the lamp; this lost energy is converted to heat.

Candela - Unit of luminous intensity, describing the intensity of a light source in a specific direction.

Candela Distribution - A curve, often on polar coordinates, illustrating the variation of luminous intensity of a lamp or luminaire in a plane through the light center.

Candlepower - A measure of luminous intensity of a light source in a specific direction, measured in candelas (see above).

CBM - Certified Ballast Manufacturers Association; the CBM label indicates that the ballast has been tested to meet ANSI specifications.

Coefficient of Utilization (CU) - The fraction of bare lamp lumens (inside a specific luminaire) that are delivered to the workplane in a specified room.

Color Rendering Index (CRI) - A scale for the effect of a light source on the color appearance of an object in comparison with the color appearance under a reference light source. Expressed on a scale from 0 to 100, where 100 is no color shift. In general, a low CRI rating indicates that the colors of objects will appear unnatural under that particular light source.

Color Temperature - The color temperature is a specification of the color appearance of a light source, relating the color to a reference source that is heated to a particular temperature, measured by the thermal unit Kelvin. The measurement can also be described as the "warmth" or "coolness" of a light source. Generally, sources below 3500K are considered "warm;" while those above 4000K are considered "cool" sources.

Compact Fluorescent - A small fluorescent lamp that is often used as an alternative to incandescent lighting. The lamp life is about 10 times longer than incandescent lamps and is 3-4 times more efficacious. Also referred to as PL, DL, CFL or BIAX lamps.

Constant Wattage (CW) Ballast - A premium type of HID ballast in which the primary and secondary coils are isolated. Considered a high performance, high loss ballast featuring excellent output regulation.

Constant Wattage Autotransformer (CWA) Ballast - A popular type of HID ballast in which the primary and secondary coils are electrically connected. Considered an appropriate balance between cost and performance.

Contrast - The relationship between the luminance (brightness) of an object and its immediate background. For example, this page is high contrast because the letters are black and the paper is white.

CRI - See Color Rendering Index.

Cut-off Angle - The angle from a fixture's vertical axis at which a reflector, louver or other shielding device cuts off direct visibility of a lamp. It is the complementary angle of the shielding angle.

CW - Designation for the cool-white halophosphor used in fluorescent lamps. Cool-white lamps have a CRI rating of 62 and a color temperature of about 4200K.

DALI - Digital Addressable Lighting Interface, which is an international standard protocol for digital control of DALI-compliant ballasts and relays using a 16-bit message format. Each DALI message includes an individual or group address as well as a command. Among the many commands available, the most common include "go to level x", "go to scene x", "set fade time x" and "set maximum light level x."

Daylight Compensation - A dimming system controlled by a photocell that reduces the output of the lamps when daylight is present. As daylight levels increase, lamp intensity decreases. An energy-saving technique used in areas with significant daylight contribution.

Diffuse - Term describing dispersed light distribution. Refers to the scattering or softening of light.

Diffuser - A translucent piece of glass or plastic sheet that shields the light source in a fixture. The light transmitted throughout the diffuser will be redirected and scattered.

Direct Glare - Glare that is produced by a direct view of light sources. Often the result of insufficiently shielded light sources. See Glare.

Downlight - A type of ceiling luminaire, usually fully recessed, where most of the light is directed downward. May feature an open reflector and/or shielding device.

Efficacy - A metric used to compare light output to energy consumption. Efficacy is measured in lumens per watt. Efficacy is similar to efficiency, but is expressed in dissimilar units. For example, if a 100W source produces 1700 lumens, then the efficacy is 17 lm/W.

Electroluminescent - A light source technology used in exit signs that provides uniform brightness, long lamp life (approximately eight years), while consuming very little energy (less than one watt per lamp).

Electronic Ballast - A ballast that uses semiconductor components to increase the frequency of fluorescent lamp operation—typically in the 20-40 kHz range. Smaller inductive components are used to provide the lamp current control. Fluorescent system efficiency is increased due to high frequency lamp operation.

Electronic Dimming Ballast - A variable output electronic fluorescent ballast.

EMI - Abbreviation for Electromagnetic Interference. High frequency interference (electrical noise) caused by electronic components or fluorescent lamps that interferes with the operation of electrical equipment. EMI is measured in micro-volts, and can be controlled by filters. Because EMI can interfere with communication devices, the Federal Communication Commission (FCC) has established limits for EMI. EMI can also be radiated; see Radio Frequency Interference.

Energy-Saving Ballast - A type of magnetic ballast designed so that the components operate more efficiently, cooler and longer than a "standard magnetic" ballast. By U.S. law, standard magnetic ballasts can no longer be manufactured.

Energy-Saving Lamp - A lower wattage lamp, generally producing fewer lumens.

Flicker - Variation in light intensity due to 60 Hz operation. Can cause eyestrain and fatigue due to stroboscopic effects.

Fluorescent Lamp - A light source consisting of a tube filled with argon, along with krypton or other inert gas. When electrical current is applied, the resulting arc emits ultraviolet radiation that excites the phosphors on the inside of the lamp wall, causing them to radiate visible light.

Footcandle (fc) - The English unit of measurement of the illuminance (or light level) on a surface. One footcandle is equal to one lumen per square foot.

Footlambert (fl) - A unit of luminance (or brightness) equivalent to the uniform emittance or reflectance of one lumen per square foot from a perfectly diffusing surface. Also, the number of footcandles striking a surface multiplied by the reflectance of that surface as seen from a given direction. See Luminance.

Glare - The effect of brightness or differences in brightness within the visual field sufficiently high to cause annoyance, discomfort or loss of visual performance.

Halogen - See Tungsten Halogen Lamp.

Harmonic Distortion - A harmonic is a sinusoidal component of a periodic wave having a frequency that is a multiple of the fundamental frequency. Harmonic distortion from lighting equipment can interfere with other appliances, as well as the operation of electric power networks. The total harmonic distortion (THD) is usually expressed in a percentage of the fundamental line current. THD for 4-foot fluorescent ballasts usually range from 5% to 25%. For compact fluorescent ballasts, THD levels can be significantly higher.

HID - Abbreviation for High Intensity Discharge. Generic term used to describe mercury vapor, metal halide, high pressure sodium and (informally) low pressure sodium light sources and luminaires.
High-Bay - Pertains to the type of lighting in an industrial application where the ceiling is 20 ft. or higher. Also describes the application itself.

High-Output - A lamp or ballast designed to operate at higher currents (800mA) and produce more light.

High Power Factor - A ballast with a 0.9 or higher rated power factor, which is achieved by using a capacitor.

High-pressure Sodium Lamp (HPS) - A high-intensity discharge (HID) lamp whose light is produced by radiation from sodium vapor (and mercury).

Hot Restart (or Hot Restrike) - A phenomenon of re-striking the arc in an HID light source after a momentary power loss. Hot restart occurs after the arc tube has cooled a sufficient amount.

IESNA - Abbreviation for Illuminating Engineering Society of North America.

Illuminance - A photometric term that quantifies light incident on a surface or plane. Illuminance is commonly referred to as light level. It is expressed as lumens per square foot (footcandles), or lumens per square meter (lux).

Indirect Glare - Glare that is produced from a reflective surface.

Instant Start - A fluorescent circuit that ignites the lamp instantly with a very high starting voltage from the ballast. Instant start lamps have single-pin bases.

Lamp Lumen Depreciation (LLD) Factor - A factor that represents the reduction of lamp lumen output over time. The factor is commonly used as a multiplier to the initial lumen rating in illuminance calculations, which compensates for the lumen depreciation. The LLD factor is a dimensionless value between 0 and 1.

Lamp Current Crest Factor - The peak current divided by the RMS current. Although the CCF may range from one to infinity, CCFs exceeding 1.7 may void lamp warranties.

Lay-in Troffer - A fluorescent fixture; usually a 2' × 4' fixture that sets or "lays" into a specific ceiling grid.

Lens - A transparent or translucent medium that alters the directional characteristics of light passing through it. Usually made of glass or acrylic.

Light Loss Factor (LLF) - A factor used in calculating illuminance after a given period of time and under given conditions. It is the product of maintenance factors, such as lamp lumen depreciation, luminaire dirt depreciation, lamp burn-outs and room surface dirt depreciation. Non-recoverable factors such as ballast factor and luminaire surface depreciation may also be multiplied. This number will be less than 1.

Light Trespass - A situation that occurs when, due to lack of adequate beam control, light from an outdoor source is distributed onto areas where the illumination is not wanted.

Load Shedding - A lighting control strategy for selectively reducing the output of light fixtures on a temporary basis as a means to reduce peak demand charges.

Louver - Grid type of optical assembly used to control light distribution from a fixture. Can range from small-cell plastic to large-cell anodized aluminum louvers used in parabolic fluorescent fixtures.

Low Power Factor - Essentially, an uncorrected ballast power factor of less than 0.90. See NPF.

Low-pressure Sodium Lamp (LPS) - A low-pressure discharge lamp in which light is produced by radiation from sodium vapor. Considered a monochromatic light source (most colors are rendered as gray).

Low-voltage Lamps - A lamp—typically compact halogen—that provides both high intensity and good color rendition. Lamp operates at 12V and requires the use of a transformer. Popular lamps are MR11, MR16 and PAR36.

Low-voltage Switch - A relay (magnetically operated switch) that permits local and remote control of lights, including centralized time clocks or computer control.

Lumen - A unit of light flow, or luminous flux. The lumen rating of a lamp is a measure of the total light output of the lamp.

Lumen Maintenance Control - An energy-saving lighting control strategy in which a photocell is used with a dimming system to provide a fixed light level over the maintenance cycle.

Luminaire - A complete lighting unit consisting of a lamp or lamps, along with the parts designed to distribute the light, hold the lamps and connect the lamps to a power source. Also called a fixture.

Luminaire Efficiency - The ratio of total lumen output of a luminaire and the lumen output of the lamps, expressed as a percentage. For example, if two luminaires use the same lamps, more light will be emitted from the fixture with the higher efficiency.

Luminance - A photometric term that quantifies brightness of a light source or of a surface that is illuminated and reflects light. It is expressed as footlamberts (English units) or candelas per square meter (metric units).

Lux - The metric unit of measure for illuminance of a surface. One lux is equal to one lumen per square meter. One lux equals 0.093 footcandles.

Maintained Footcandles - Footcandles calculated through application of light loss factors.

Mercury Vapor Lamp - A type of HID lamp in which the major portion of the light is produced by radiation from mercury vapor. Emits a blue-green cast of light. Available in clear and phosphor-coated lamps.

Metal Halide Lamp - A type of HID lamp in which the major portion of the light is produced by radiation of metal halide and mercury vapors in the arc tube. Available in clear and phosphor-coated lamps.

MR16 - A low-voltage quartz reflector lamp, only 2 inches in diameter. Typically the lamp and reflector are one unit, which directs a sharp, precise beam of light.

NADIR - A reference direction directly below a luminaire, or "straight down" (0 degree angle).

NPF, Normal Power Factor - A ballast/lamp combination in which no components (e.g. capacitors) have been added to correct the power factor, hence normal (essentially low) power factor (typically 0.5 or 50%).

Occupancy Sensor - Control device that turns lights off after the space becomes unoccupied. May be ultrasonic, infrared, combination or other type.

Optics - A term referring to the components of a light fixture (such as reflectors, refractors, lenses, louvers, etc.) or to the light emitting or light-controlling performance of a fixture.

PAR Lamp - A Parabolic Aluminized Reflector lamp. An incandescent, metal halide or compact fluorescent lamp used to redirect light from the source using a parabolic reflector. Lamps are available with flood or spot distributions.

PAR36 Lamp - A PAR lamp that is 36 one-eighths of an inch in diameter with a parabolic shaped reflector. See PAR Lamp above.

Parabolic Luminaire - A popular type of fluorescent fixture which has a louver composed of aluminum baffles that are curved in a parabolic shape. The resultant light distribution produced by this shape provides reduced glare, better light control, and is considered to have greater aesthetic appeal.

Paracube - A metallic coated plastic louver made up of small squares. Often used to replace the lens in an installed troffer to enhance its appearance. The paracube is visually comfortable, but the luminaire efficiency is lowered. Also used in rooms with computer screens because of their glare-reducing qualities.

Photocell - A light sensing device used to control luminaires and dimmers in response to detected light levels.

Photometric Report - A photometric report is a set of printed data describing the light distribution, efficiency and zonal lumen output of a luminaire. This report is generated from laboratory testing.

Potting - The filler material used in some magnetic and electronic ballasts. Many ballasts are filled with tar or plastic material to dissipate heat and noise from the electrical components.

Power Factor - The ratio of AC wattage through a device to AC volts x amps of the device. A device such as a ballast that measures 120 volts, 1 amp and 60 watts has a power factor of 50% (volts x amps = 120 VA, therefore 60 watts ÷ 120 VA = 0.5) Some utilities charge their customers for low power factor systems.

Preheat - A type of lamp/ballast circuit that uses a separate starter to heat up a fluorescent lamp before high voltage is applied to start the lamp.

Quad-tube Lamp - A compact fluorescent lamp with a double twin-tube configuration.

Radio Frequency Interference (RFI) - Interference to the radio frequency band caused by other high frequency equipment or devices in the immediate area. Fluorescent lighting systems generate RFI.

Rapid Start (RS) - The most popular fluorescent lamp/ballast combination used today. This ballast is designed to quickly and efficiently preheat lamp cathodes to start the lamp. Uses a "bi-pin" base.

Reflectance - The ratio of light reflected from a surface to the light incident on the surface. Reflectances are often used for lighting calculations. The reflectance of a dark carpet is around 20%, and a clean white wall is roughly 50-60%.

Reflector - The part of a light fixture that shrouds the lamps and redirects some of the light emitted from the lamp.

Refractor - A device used to redirect the light output from a source, primarily by bending the waves of light.

Recessed - The term used to describe the door-frame of a troffer where the lens or louver lies above the surface of the ceiling.

Regulation - The ability of a ballast to hold constant (or nearly constant) the light output (ballast output watts) during fluctuations in the input voltage. Normally specified as +/- percent change in output compared to +/- percent change in input.

Relay - A device that performs the actual on or off switching of an electrical load due to small changes in current or voltages. Examples: low voltage relay and solid state relay.

Retrofit - Refers to upgrading a fixture, room, building, etc., by installing new parts or equipment.

Room Cavity Ratio - Room Cavity Ratio (RCR) is a ratio of room dimensions used to quantify how light will interact with room surfaces. A factor used in illuminance calculations.

Room Surface Dirt Depreciation - A factor used in illumination calculations which represents the fractional loss of task illuminance due to dirt on the room surface. This number is always less than 1.

Scheduling - An energy-saving lighting control strategy for dimming or switching lighting systems during hours when a building space is unoccupied or occupied by individuals with less stringent lighting requirements.

Semi-Specular - Term describing the light reflection characteristics of a material. Some of the light is reflected directionally, with some amount of scatter.

Shielding Angle - The angle measured from the ceiling plane to the line of sight where the bare lamp in a luminaire becomes visible. Higher shielding angles reduce direct glare. It is the complementary angle of the cutoff angle. See Cut-Off Angle.

Snap-Back - The potential or real situation where an energy- efficiency upgrade could be replaced with the original type of equipment. Installations that are subject to snap-back are not permanent.

Spacing Criterion - The maximum recommended distance that interior fixtures should be spaced to ensure uniform illumination on the workplane. The height of the luminaire above the workplane multiplied by the spacing criterion equals the recommended maximum center-to-center luminaire spacing.

Specular - Mirrored or polished surface. The angle of reflection is equal to the angle of incidence. This word is used to describe the finish of the material used in some louvers and reflectors.

Stroboscopic Effect - Condition where rotating machinery or other rapidly moving objects appear to be standing still due to the alternating current supplied to light sources. Sometimes called "strobe effect."

T12 Lamp - Industry standard for a fluorescent lamp that is 12 one-eighths (1-1/2 inches) in diameter. Other typical sizes include T10 (1-1/4 inches) and T8 (1 inch) lamps.

Tandem Wiring - A wiring option in which a ballast is shared by two or more luminaires. This reduces material and energy costs. Also called "master-slave" wiring.

Task Lighting - The lighting, or amount of light, that falls on a given visual task.

THD - See Harmonic Distortion.

Thermal Factor - A factor used in lighting calculations that compensates for the change in light output or wattage of a fluorescent lamp due to a change in bulb wall temperature. It is applied when the lamp-ballast combination under consideration is different from that used in the photometric tests.

Trigger Start Ballast - Type of ballast commonly used with 15W and 20W straight fluorescent lamps.

Troffer - The term used to refer to a recessed fluorescent light fixture (combination of trough and coffer).

Tungsten Halogen Lamps - A gas-filled tungsten filament incandescent lamp with a lamp envelope made of quartz to withstand the high temperature. This lamp contains a certain proportion of halogens, namely iodine, chlorine, bromine and fluorine that slows down the evaporation of the tungsten. Also commonly referred to as a quartz lamp.

Tuning - An energy-saving lighting control strategy in which the light output of an individual fixture or group of fixtures is adjusted to provide the correct amount of light for a local task.

Twin-tube - See Compact Fluorescent Lamp.

Ultraviolet (UV) - Invisible radiation that is shorter in wavelength and higher in frequency than visible violet light (literally beyond the violet light).

Underwriters Laboratories (UL) Inc. - An independent organization whose responsibilities include rigorous testing of electrical products. When products pass these tests, they can be labeled (and advertised) as "UL listed." UL tests for product safety only.

Vandal-resistant - Fixtures with rugged housings, break-resistant type shielding and tamperproof screws.

VCP, Visual Comfort Probability - A rating system for evaluating direct discomfort glare. This method is a subjective evaluation of visual comfort expressed as the percent of occupants of a space who will not be bothered by direct glare. VCP takes into account luminaire luminances at different angles of view, luminaire size, room size, luminaire mounting height, illuminance and room surface reflectivity. VCP tables are often provided as part of photometric reports.

Veiling Reflection - Also known as a type of reflected glare, a reflection of a light source that partially or totally obscures details by reducing the contrast between task details and their background.

Very High Output - A fluorescent lamp that operates at a "very high" current (1500mA), producing more light output than a "high output" lamp (800mA) or standard output lamp (430mA).

Volt - The standard unit of measurement for electrical potential. It defines the "force" or "pressure" of electricity.

Voltage - The difference in electrical potential between two points of an electrical circuit.

Wall Washer - Term used to describe the luminaires designed to illuminate vertical surfaces.

Watt (W) - The unit for measuring electrical power. It defines the rate of energy consumption by an electrical device when it is in operation. The energy cost of operating an electrical device is determined by its wattage times the hours of use. In single phase circuits, it is related to volts and amps by the formula: Volts x Amps x Power Factor = Watts. (Note: For AC circuits, power factor must be included.)

Workplane - The level at which work is done and at which illuminance is specified and measured. For office applications, this is typically a horizontal plane 30 inches above the floor (desk height).

WW, Warm White - Designation for the warm-white halophosphor used in fluorescent lamps. Warm-white lamps have a CRI rating of 53 and a color temperature of about 3000K.

Zenith - The direction directly above the luminaire (180 degree angle).

Index

Printed in the United States
by Baker & Taylor Publisher Services